WILDLY
SUCCESSFUL
PLANTS

WILDLY SUCCESSFUL PLANTS

A Handbook of North American Weeds

by **LAWRENCE J. CROCKETT**

Illustrations by

Joanne Bradley

Collier Books

A DIVISION OF MACMILLAN CO., INC.

NEW YORK

Collier Macmillan Publishers

LONDON

Macmillan Publishing Co., Inc.
866 Third Avenue, New York, N.Y. 10022
Collier Macmillan Canada, Ltd.

Library of Congress Cataloging in Publication Data

Crockett, Lawrence J
   Wildly successful plants.

   Bibliography: p.
   1. Weeds—United States—Identification. 2. Weeds—
Canada—Identification. I. Title.
SB612.A2C72   581.6′5   76-54687
ISBN 0-02-528850-4
ISBN 0-02-062600-2 pbk.

First Printing 1977

Designed by Jack Meserole

WILDLY SUCCESSFUL PLANTS is also published in a hardcover edition
by Macmillan Publishing Co., Inc.

Printed in the United States of America

*This book is dedicated to my teacher and friend,
the late* DR. EDWIN B. MATZKE, *Professor of Botany,
Columbia University in the City of New York*

Contents

Preface, xi

All You Need to Know about Weeds, 1

Some Basic Facts about Flowering Plants, 6

How to Identify the Weeds in This Book, 18

FLOWERING PLANTS

HERBACEOUS PLANTS
Monocotyledonous Species

HERBACEOUS PLANTS
Dicotyledonous Species

Green-Flowered Species

Yellow-Green–Flowered Species

Yellow-Flowered Species

Orange-Flowered Species

White-Flowered Species

Pink- or Red-Flowered Species

Blue-Flowered Species

Purple-Flowered Species

 NONFLOWERING PLANTS

Plants without Flowers

APPENDICES

Preface

In 1865, as the Civil War came to a close, some 85 percent of our citizens were living in farming areas where they grew the food for themselves and for the remaining 15 percent of the population—our "city folk." By the mid-1970s, just a century later, a silent revolution had occurred: a revolution with no bloodletting, no guillotines, no tea parties or tennis court oaths later violated. Now only 7 percent (or fewer) of our population is feeding the rest of us—and a large part of the underdeveloped world that is unable to feed itself. That same 7 percent has already created enough surplus to export grain to the supposedly advanced Soviet Union, temporarily incapable of feeding its citizens though some 60 percent of them are still directly involved in agriculture. So much for the socialist paradise!

In light of the huge population shift from rural areas to the cities of America, it is easy to see why our major cities are in trouble today. That shift has also changed our general opinion of what constitutes a threatening weedy species of plant. "Weediness" today summons the image of an unwanted, invading plant in the garden, on the lawn, or along the roadside. A century ago most Americans were conscious of, and concerned about, the invasion by weedy species of their crop lands.

While weedy plants may seem bothersome in one way or another, these sometimes unwanted plants have played a very large role in humankind's progress toward a greater civilization. Some weedy species are, indeed, an integral part of our civilization. It is very likely that our basic agricultural plants began as invading weeds thousands of years ago. Slowly (and probably accidentally) the primitive early farmers observed that the invader offered a far better food source. The discovery of this utilitarian food source made greater civilization possible, i.e., more efficient agriculture gave humanity time away from the plow, or the leisure time in which to create civilization. Perhaps the ancient farmers learned what they did because they were lazy, too lazy to do a good job of weeding their plots. The first and the last contribution of the lazy to civilization!

Weeds are of interest not only because of their contribution to civilization, but also because they are plants, and plants are, in themselves, fascinating organisms. To some, animals are more interesting because they move (human beings probably identify more easily with organisms or creatures that move than with those that do not). Yet it is stationary plants that feed all animals and make their motion possible.

Plants do not move because, unlike animals, they make their own food. Through the process of photosynthesis in their chlorophyll-containing cells, plants manufacture *all the food in the world*. Even if you eat only steak, what you are eating is but modified grass. Isaiah, untutored in modern science, showed he knew this when he wrote, "All flesh is grass." Not only do photosynthetic plants give us all our food, but also our wood (the cellulose walls around their cells), the oxygen we breathe (a waste product of photosynthesis), the oil we still have not learned to consume wisely (their dead bodies from eons ago), coal (again, their dead bodies), and an impressive part of our artistic stimulation (plants are beautiful).

Therefore in this volume I have tried to show that plants are interesting both in themselves and as part of the larger picture of civilization. The illustrator-artist Joanne Bradley has shown the beauty, while clearly delineating the structure, of the weedy species chosen for description in this book. The Contessa Giovanna, as she wanders from figure to figure in her ample garden hat, shows the reader, at a glance, the overall size and general growth habit of each weed, and we are indebted to her for this.

For a reader not trained in the ways and habits of plant classification, or taxonomy, for anyone not recently exposed to botany, the identification of plants can be difficult and sometimes frustrating. Many special descriptive adjectives and nouns are necessary for the proper use of special taxonomical systems, and many of these terms have several shades of meaning not immediately apparent to the novice. The reader with no field experience, when reading that a certain twig is "stout," will wonder—just how stout is "stout?"

Some popular books about weeds do not include a system for their identification; others include very complex systems. I have chosen to make up a "system," based on the plants in this book, that leans considerably on everyday terms *and* a limited number of specialized botanical terms (it would truly be impossible not to use any specialized terms at all). Once the first three chapters are digested, the reader should have no difficulty grasping the botanical terms that are later used in identifying a particular weedy plant. These chapters also give the reader some idea of how plants, in general, are put together. Throughout the remainder of the book, each description is preceded by the family to which the weed belongs. A glossary at the end of the book should also prove useful. There is absolutely nothing wrong with turning directly to the drawings and trying to identify a weed by comparing it with the drawing. Possibly best of all, just read the book with not a thought, for the nonce, of identifying any particular weed. Read about these plants for their own sakes—and yours.

WILDLY
SUCCESSFUL
PLANTS

All You Need to Know about Weeds

There are no weeds in nature, just as there are no peasants. Civilization and cultivation have created both. It is obvious that botanists have had some difficulty defining the concept "weed." In general, they have come to the conclusion that a weed is any plant found growing rapidly and abundantly in a place where it is not wanted. Perhaps when the ancient farmers cleared a small piece of land, planted a crop, and then had to fight some invading plant species, that invader became the first weed. Or was the first weed a plant that, because of its rapid rank growth, contested with men the entrance to their first cave homes?

There are no records to tell us when human beings first confronted what they came to call weeds. Records of weedy plants, even from peoples who kept good records (such as the ancient Egyptians), are scanty and apparently no early civilization had a clear-cut word for a weed.

Our word "weed" descends from Middle English (by way of, and influenced by, Flemish, Anglo-Saxon, and Frisian), but the best etymologists think that the origin of the word has been lost to us. Some believe it is related to the Old English word "woad," applied to the leguminous plant that used to supply a valuable blue dye to the English, but this is by no means a clear fact.

If a weed is defined only by its unwantedness, much is being overlooked by the definer. Let us consider the most hated weed plant of the suburban lawn grower: crabgrass. *Digitaria sanguinalis* (p. 85) is the botanist's name for crabgrass, but in English or Latin this plant is the enemy of the householder and the mere mention of its name will bring blood to the whites of the eyes of suburban gentlemen. Yet the strawberry grower puts the same grass to good use (p. 85), and so does the Southern farmer, who cultivates crabgrass for forage and pasturage. So one man's plant enemy is another's good friend. There are many, many weeds in this category.

Almost all the definitions of "weed" depend on the opinion of the one who is doing the defining. Certain characteristics of weedy species are more personally important to the definer than to others, and therefore definitions that include a number of different attributes of weedy species have arisen. These have included: their danger to man or animals; their rank, rapid, wild growth; their successful growth, unbidden, unsown, and uncultivated; their usually aggressive and competitive growth habits; their insistence on growing where they are not wanted; their persistence and resistance to control or eradication; their ability to reproduce themselves rapidly both by seeds and by vegetable means; their penchant for making the landscape less attractive to the human eye. From this list it is easy to see that man is very personally involved in his definition of the concept "weed" and that a number of his reasons for disliking weedy species are psychologically colored.

Not all those who have written on the subject have dealt with weeds as unwanted or "evil" plants. A few voices have been raised in their favor, most prompted by the ecological point of view, i.e., the disposition to take the plants as plants.

A. H. Bunting said in 1960, "Weeds are pioneers of secondary succession of which the weedy arable field is the special case." F. C. King wrote, "Weeds have always been condemned without a fair trial," and Joseph Cocannouer, who has written popularly and favorably about weeds, states, "This thing of considering all weeds as bad is nonsensical!"

There is considerable evidence that many of our major agricultural species, e.g., wheat, maize (corn), rye, and rice, entered the field of agriculture as weeds, growing alongside other plants, at the time considered more useful. Lamb's-quarters [*Chenopodium album*, p. 113], one of the most common weedy flowering plant species, was apparently a plant grown by our ancestors for food. Both its leaves and seeds are edible. When lamb's-quarters was an important food plant, the ancestors of wheat, maize, or rice (long since become our major agricultural crop plants) may, indeed, have been growing along the borders of cultivated *Chenopodium* patches, or invaded those patches and had to be removed as "weeds." Perhaps by accident (a lazy early farmer?) it was discovered that the seeds of grasses were even more valuable than those of lamb's-quarters.

The gardener or preserver of the lawn or any other nontechnically trained plant lover cannot be expected to become versed in ecology or ethnobotany before trying to understand the nature of weeds, but he should be aware of some interesting facts about weeds.

By and large humanity has itself to blame for the weedy species that follow in its steps. Have you ever noticed that the most favored habitats of weeds are open, sunny places? Few weeds are tolerant of shade. It was man who opened up the sunny spots where weeds are happiest. Certainly there were open spots when the forests of the United States were intact, but these had been open for ages and the plants in them had attained a definite ecological balance. Few open areas in the United States have this balance today. Man tore down the forests, cut swathes through grassland, forest, and other long-established ecological situations, and cut open pathways for roads and railroads across the entire nation. His clothes, vehicles, and animals all carried seeds and liberally scattered them as he himself spread across the continent.

This is not to suggest that only man has opened areas suddenly, cleared large patches of the forest or grassland of all their established vegetation. Glaciation, fires, floods, and other natural disturbances have occurred in the past, and will occur in the future, but man is the greatest and most persistent disturber of established plant growth patterns. Gardens are not natural; farms are not natural; lawns are not natural. No, man is the chief disrupter of long-established ecological plant relationships.

Man and the weedy plants are closer friends than he generally seems to know; indeed, his civilization may depend on that friendship. The human being and the weed are wed, though to listen to most people talk about weeds, one would think the relationship was out of *Who's Afraid of Virginia Woolf?*

It is interesting to note how few weeds are native to the United States. When the white man began to colonize this country, he brought with him familiar vegetable plants, kitchen herbs to spice his foods, medicinal plants to protect him in the frightening New World, and garden ornamentals to remind him of the old sod and its floral beauties. After he was here a while, he found that his adopted land had plants that could serve all these purposes and he included these in his farm and garden. The plants he had brought with him made their escape over fence or wall and, as species happiest in wide open spaces, became naturalized weeds, such as *Linaria vulgaris* (butter-and-eggs, p. 139).

Other species came unbidden and unsuspected. They came as seeds attached to the clothes of arriving colonists or to the fur of imported farm or domestic animals. Soil used as ballast in English ships in the eighteenth century was dropped in our Eastern port towns when rich cargoes were taken up in its place for the return trip to the mother country. From that ballast hundreds of new plant species were introduced to this country.

The noted weed expert J. M. Fogg has found that 14 percent of the plants listed in Gray's *Manual of Botany* are introduced plants. Somewhat more than half the foreign species came from Europe, and Eurasia and eastern Asia together contributed about a third as many as Europe. Tropical America has donated less than one hundred species.

Certain families of flowering plants predominate in the numbers of weedy species they contain. They are, in order of the numbers of weeds they have produced, the composite, grass, mustard, legume, pink, mint, and snapdragon families. The overall majority of nonwoody plants, and all the families listed above, are relatively recently evolved: the composite, mint, and snapdragon families are thought to be among the most recently evolved plants on earth. Weeds deserve some credit for their pioneering ability; they certainly suggest genetic hardiness.

In fact, the aggressiveness of introduced weeds is a threat. It is their very vigor that drives out not only our garden plants but also many of our native wild species. Thus we may not define a weed exclusively as a plant we do not want where it grows in garden, lawn or landscaped highway. Both native and introduced weedy species conform to this definition, but introduced species may, in addition, drive out our native wild species, denying them their former habitats and ecological niches. The Japanese honeysuckle [*Lonicera japonica*, p. 49] will usurp territory at an amazingly rapid rate and in all directions as soon as it has set itself up in a new location. It will then proceed to crowd out most native plants, thus reducing the variety of plant life in the area. The purple loosestrife [*Lythrum salicaria*, p. 239] is guilty of the same process in wetter areas. Both honeysuckle and purple loosestrife are lovely to look at when cultivated or controlled, but uncontrolled, they endanger our native wild plants and become weeds.

Weed Dispersal

Surely when faced with a yardful of rapidly growing weeds you have thought, what have weeds got that cultivated plants seem to lack? This question is partly answerable and partly a mystery to plant scientists. Weed species generally grow very rapidly after germination from seed, and this gives them an immediate advantage. Their roots get to the moisture-laden soil before those of the cultivated species, and their fast-growing stems and leaves deprive of sunlight those plants that are growing less rapidly around them. Garden ornamentals are often plants that are happiest in bright sunlight, i.e., intolerant of shade. Once the weed species has raised its leaves above the young cultivated plant, the latter's time is almost up; at the very least, its growth will be severely inhibited. Why roots and stems of weeds grow faster, why their roots put out more root hairs, why their germination from seed is more rapid, is fundamentally impossible to explain except in very general terms such as "genetic vigor."

On a particularly dry, hot August afternoon when the hydrangea leaves are hanging down like the ears of a beagle and the cultivated species can almost be heard to pant, take a good look at weedy plants. Chances are they won't look nearly as heat-worn as your more delicate species. The reason is that the weedy species have grown a greater number of root hairs and pushed their roots deeper and faster into the soil than have the cultivated plants. Their vigor is obvious.

Many, many weedy species produce huge numbers of seeds. This is the case with members of the large tribe of composites. Composites, e.g., the daisy (p. 193), do not produce a single flower, though the eye sees only a single flower. Rather, they have a head of many flowers all packed together, giving the appearance of a single flower. There are two types of flowers in the daisy head: a ray flower, produced at the edge of the head, and a disc flower, yellow in color, of which the center yellow disc of the daisy is composed. Since the flowers of the head mature in a spiral from the outside toward the center of the head, seeds set over a long period of time. Other types of plants usually present but one or a few flowers at a time to the pollinating insect. Their seeds are shed at once on the maturation or opening of the fruit, or shortly after the fall of the fruit.

Some weeds are provided with special dispersal mechanisms; one such is the dandelion. Each seed is provided with its private parachute (a modified calyx) to carry it through the air (p. 135). Other weed species produce fruits that are barbed or covered with teeth that stick to the fur of passing animals (or to the clothes of passing human beings) and are then carried great distances before being brushed away or scratched off.

The seeds of weedy species tend to be long-lived. In 1879 an American botanist, Dr. W. J. Beal, at the Michigan Agricultural Experimentation Station in East Lansing, tested the longevity of weed seeds in an experiment that is still running. He buried the seeds of twenty common weedy species in containers. At intervals of five years until 1920, and at ten years after 1920, germination tests were run on the seeds. Thirty years after the seeds were buried, about 50 percent of the species tested still had viable seeds. In 1960, some eighty years after the seeds were buried, seeds of three weeds were still alive, and germinated. They were weeds described later in the book: curly dock [*Rumex crispus*, p. 123], evening primrose [*Oenothera biennis*, p. 143], and moth mullein [*Verbascum blattaria*, p. 151]. It is expected that seeds of the three will still be alive when the centennial of the experiment is celebrated in 1979.

Not only do weeds produce huge numbers of seeds that are easily dispersed, but many of them also produce underground stems or rhizomes (sometimes called rootstocks). Rhizomes are the bane of the gardener. Indeed, *perennial* weeds (those that live from year to year) produce both aerial and underground stems and can spread even if they never get the chance to produce a single seed. Weeds with rhizomes are the most difficult to eradicate. The aerial stems may be cut away or chemically poisoned, but if the rootstock is left undamaged, the plant will continue to produce aerial shoots.

The rootstock of some weeds seems to delight in being fragmented, each piece giving rise to a new, complete plant. Cutting back the aerial stems, which make the foods stored in the rhizome, will eventually starve the underground parts, but this is a tedious, time-consuming task. Chemical attack on the rootstock is not always easy or successful either.

The aerial stem may also be involved in the direct spread of a weedy species. Aerial stems of many species take root wherever they touch the soil, especially at their nodes (where leaves appear on the stem). Essentially, a new plant is formed, which itself then sends out runners. If such a plant also produces abundant seed, it is a difficult one to beat!

Annual weeds live but one year and then set seed. Death follows soon after. These are the easiest weeds to defeat if you cut the stem before seed set. *Winter annuals* germinate in the fall, producing a root and a few leaves. Then, having a head start the next spring, they produce their flowers and seeds early.

Biennials live for two years, usually producing a rosette of leaves during the first year. The rosette is produced on the surface of the soil. A flowering stalk arises during the second year of growth. If these plants are torn out or killed with chemicals during their first year, they will no longer be a gardening headache. But if you let them set seed, though they themselves die at the end of the second season of growth, your problems will have multiplied many times and will lie completely hidden until germination time the next year and for many years to come.

In the pages to follow there will be much information about weeds and their ways.

Some Basic Facts about Flowering Plants

All but two of the weedy species presented in this book are flowering plants or *Angiospermae*. Therefore the contents of this chapter may help you to understand the nature and structure of *all* flowering plants, weedy as well as nonweedy species.

Flowering plants usually possess an aboveground *shoot system* composed of stem, leaf, axillary or lateral buds, eventually a flower or flowers, and, at the very tip of the stem, a shoot meristem or growing point. Below ground level is a *root system* of which every branch is tipped by a root meristem or growing point.

If the plant is a *perennial*, i.e., grows from year to year for more than two years, it will have within its stem and root a means for expanding laterally or in girth. Here will be found special growth tissues called *cambia*. There are two such cambia within older roots and stems. One is the *vascular cambium*, which produces new conducting tissues, *xylem*, by taking water and minerals through the root and stem (usually in an upward direction), and *phloem* (pronounced as if spelled *flome*), by carrying manufactured carbohydrates and other substances from the leaves where they are made to the other parts of the plant (usually in a downward direction). The other is the *cork cambium*, which gives rise to new storage tissues in stem and root (called cortex), and to the most external tissue (the cork), which protects the outside of older roots and stems.

Thus when you point to a tree and say, "Oh, look at that interesting bark!" you are not entirely correct botanically, for what you are indicating is not the bark but the cork. Scientists must define everything very carefully, and to plant scientists or botanists, bark includes all tissues from the internal tissue, the vascular cambium, out to the cork, or external tissues.

The root system, too, has apical meristems or growing points on the tips of all its roots. Cells, constantly formed within the meristems, later mature as various root tissues (xylem and phloem and cortex are examples), and the increase in cells pushes the root down through the soil, while the shoot meristems keep the shoot pushing its way through the air. Hormone systems keep each organ growing in the correct (and favorable) direction. Thus the vertical growth of the plant is accomplished by its meristems (root and shoot). Growth accomplished by the meristems at the tips of the root is known as *primary growth*. Since plants called annuals live for only one growing season and do not develop cambia, they are produced entirely by primary growth. So, too, are biennials, which live for only two growing seasons. The pigweed [*Chenopodium album*, p. 113] is an annual, and mullein [*Verbascum thapsus*, p. 149] is a biennial.

While annuals are composed fundamentally of tissues produced during primary growth, perennials include an additional kind of growth. Not only do they grow by means of their shoot and root meristems, but they also develop cambia in both systems and thus grow laterally as well. Growth produced by cambia is known as secondary growth. Perennials have functioning cambia by the end of their first year of growth. The tree-that-grows-in-Brooklyn [*Ailanthus altissima*, p. 37] is a perennial.

Within the perennial plant it is the annual increase of xylem tissue (which forms as a ring of tissue) that permits botanists to determine the age of the tree by counting the rings. *Annual rings* also give clues to the ecological conditions through which the tree has passed during its years of growth. Anthropologists now use tree rings as part of their technique for determining the age of villages in the American Southwest, a method first worked out by Professor A. E. Douglass of the University of Arizona.

One more word about growing points of the stems. This region has been called a region of *continued embryology* and it proves plant growth and development are clearly different from what is found among animals. In a five-thousand-year-old bristlecone pine [*Pinus aristata*—definitely not a weed] the cells of the growing points of root and shoot are the direct descendants of those cells alive five thousand years ago at the tips of the oldest branches and roots of this botanical Methuselah.

The stem of the perennial plant remains all winter and does not die back to the ground. While its leaves usually fall away with the approach of autumn, left behind is the whole woody branch and its buds, both axillary and terminal (if a terminal bud is present); leaf scars, where last summer's leaves were attached; and cork on the outside, protecting and preventing water from escaping and bacteria from entering. The cork provides another help to the living tissues within the stem (cork cells are, by the way, dead). Small regions of looser cork cells, appearing as tiny clearer areas of varying shapes—the *lenticels*—permit the exchange of gases so necessary to maintaining life. Through the lenticels pass carbon dioxide and oxygen; the former leaves and the latter enters. Both are involved in respiration. Carbon dioxide is a waste product of this energy-storing process and oxygen is consumed. All living cells, in the plant stem or anywhere (in the human body, for example), must carry on respiration or perish. Plants, unlike animals, also carry on photosynthesis. Respiration is a continuous action; photosynthesis is carried on only in the light.

The twig may be capped by a *terminal bud,* or the last axillary bud formed during the previous growing season will take over and act as a terminal bud, as in the tree-that-grows-in-Brooklyn [*Ailanthus altissima,* p. 37]. This kind of bud is called a *pseudoterminal bud.*

When, in the spring, new growth begins once more, the bud scales around the terminal bud will fall and the contents of the bud will expand. A line is left about the stem marking where the bud scales were attached. Bud scales are modified leaves. Counting successive lines of bud scale scars is one way of determining the age of a branch, just as the more internal annual rings of secondary xylem can be used to determine the age of a tree.

Some Modifications of the Stem

Not all stems grow erect. The *rhizome* is an underground stem that sends up leaves from its "dorsal" surface and roots from its "ventral" surface. In temperate climes ferns are characterized by rhizomes, though in more tropical climates some ferns grow as erect, tall trees. Many flowering plants also produce rhizomes, and they are quite common among the grasses. Fragmentation of the rhizome produces a number of individual plants and is thus a form of asexual reproduction.

A thickened portion of a rhizome is a *tuber*. The tuber you know best is the white potato (not originally from Ireland but from Peru), which is a stem rather than a root. Look at a good-sized potato and note the cork or dark "skin" on the outside; the "eyes," which are actually axillary buds (and can, therefore, produce a new plant if roots can be induced to grow); and, possibly, a thin stalk protruding from one end of the potato, which represents all that is left of the normal part of the underground stem not filled with stored food, as is the tuber.

Some underground stems are nubbinlike structures that are called *corms* or solid bulbs. Crocus is grown from a corm. The true *bulb*, however, while it contains a short, thickened stem, is fundamentally composed of overlapping leaves filled with stored food. Take a whole onion and cut through it longitudinally, i.e., from the leafy part on top to the flat root-bearing part below, and look inside. Here the food-rich swollen leaves are obvious, and the small, flat stem to which they are attached (and to which the roots are attached) is also rather easy to see.

Bulbils are small editions of bulbs, but they form far above the ground in either axillary bud regions or within the inflorescence. The latter method of formation is common among onion relatives. Wild garlic [*Allium vineale*, p. 73], because of its bulbil formation, can be the bane of the lawn keeper or preserver of fine pastures. When the bulbil falls to the ground, it is ready to begin growth as soon as its roots dig into the soil.

Armaments Seen on Stems

Poetry is filled with roses that have thorns. Unfortunately for poetic accuracy, the rose does not produce a thorn! Rather the skin-stabbing structure on this beautiful flowering plant is a *prickle* (p. 47). Scientists must define everything very carefully: a true *thorn* is a modified stem, which the skin-stabbing structure of the rose is not. In a true thorn an axillary bud grows out, gradually hardening, until finally its apex points up. Thus the thorn is a branch that has become tough and pointed. Thorns are very hard to break off because they are so intimately connected to the main trunk or branch. The rose prickle is an outgrowth of the epidermal tissues of the stem, as are all prickles; however, I prefer that the poet continue to use thorn (his license) since so few lovely words rhyme with prickle! A third armament is found on stems. This one, which can be noted on barberry (p. 45), is a *spine* and actually represents a modified leaf.

General Positions of Stems

The shoot system may grow *erect* (standing upright), as most are observed to do, bend over and lean in various degrees (p. 46), or lie *prostrate* on the ground. This last condition is also called *procumbent—decumbent* if the stem bows nearly to the ground. A *creeping* or *repent* stem lies on the ground but produces adventitious roots (roots that appear where normally roots do not) at its nodes. (Weeds that can do this are a very special nuisance to the gardener.) If the stem can't support itself at all, it may climb as a *vine* and support itself entirely by clinging. Some vines twine about their supports—for example, the morning glory [*Convolvulus sepium*, p. 51]; some use specialized roots to hold onto the support—for example, ivy, poison ivy [*Rhus radicans*, p. 61]; and some produce special structures called *tendrils*—for example, the fox grape [*Vitis labrusca*, p. 55]. However, the tendrils of the garden pea arise as leaflets and those of the wild cucumber [*Echinocystis lobata*, p. 59] are whole modified leaves.

Where Leaves Are Borne on Stems

Along the stem, leaves are borne at *nodes*, and the portion of the stem between the nodes is called an *internode*. Leaves are arranged on stems oppositely if the two leaves face each other, or in a circle if there are three or more leaves at a node (see *Equisetum arvense*, horsetail, p. 243). In the vast majority of plants the leaves are borne alternately on the stems (see *Ailanthus*, p. 37). In this case the leaves are arranged in such a way that each leaf is alone at a node and at a different level from all others. Thus the leaves form a spiral about the stem.

The Green Leaf and Its Function

Leaves are the photosynthetic organs of green plants. Photosynthesis is a biochemical process that occurs only in the presence of light and chlorophyll within the chloroplast. In the process carbon dioxide, taken from the environment through the stomates (small aperatures in the surface of leaves that may be opened and closed by special mechanisms), and water molecules, absorbed by the root and transported through the xylem, are reacted upon and changed within the chloroplast so that a carbohydrate and oxygen are manufactured. The oxygen is released as a waste product; our earthly oxygen blanket, upon which all animal life depends, gets most of its oxygen from this complex process! From the manufactured carbohydrate, say, glucose, the plant is capable of making all the various nutrients we need to live—fats, sugars, amino acids, vitamins—and they are available to us from no other source. If you dine on meat, you are dining on modified grass and nothing more!

Therefore, we can say all animal life depends absolutely on the photosynthetic process. If tomorrow the process were to fail globally, you and I and all other animals that live on land or in water would be doomed. All animals are fundamentally parasitic on all photosynthetic plants. Everything humankind knows and loves—our whole civilization—depends in the last analysis on the continuation of the photosynthetic process on land and in the waters of our planet. The time has come to give more thought to this: green or photosynthetic plants are the keystone of the food chains of this planet earth.

The Structure of the Leaf

A great number of leaves are composed of a flat, expanded portion called the *blade*, a region of attachment to the stem called a *petiole*, and in the upper angle formed by the petiole and the stem's meeting an *axillary bud*. Indeed, to be a leaf the structure must have a bud in its axil. In members of the Dicotyledonae (flowering plants are divided into two large subgroups of which the Dicotyledonae, or in its shortened form, dicot, is one) the leaf meets the stem by means of the petiole itself, usually a thin, sticklike portion of the leaf. In the Monocotyledonae (or monocot) petioles are usually lacking and here the leaf base wraps itself around the stem or clasps it. This condition is known as *clasping* or *sessile*. In the leaf of the dicot, leaf blade veins can be seen running this way and that. This condition is known as *netted venation* or *reticulate venation*. Within the leaf of the monocot the venation is quite different. The veins run from the tip of the leaf to the base and are *parallel*. The shape of the monocot leaf is usually like that of a sword, i.e., long, thin, and tapering from the base to the tip, while a typical dicot leaf is elliptical. There are, however, many exceptions and leaf shape varies tremendously among the dicots.

The manufactured foods are shipped from the cells in which they are made (chlorophyllous cells) through leaf veins to veins in the petiole, and then into the veins in the stem and to other parts of the plant that are not photosynthetic. The plant usually makes excess food and this is shipped to storage cells. Much food is also stored in seeds. The manufactured foods travel in the phloem tissue.

With a few exceptions, members of the monocots do not produce cambia in their stems and roots and thus do not show secondary growth. The palms are one exception. Thus, no matter how thick the stem may appear, e.g., as in the "trunk" of the banana "tree," it is composed of a relatively thin stem that is made fat with overlapping, clasping leaf bases. Tulip, lily, crocus, onion, gladiolus, and grass are common monocots.

The Axillary Bud

Within the axil of the leaf is found the lateral bud. In perennials, when the leaf falls at the approach of winter, the axillary buds usually become more easy to see than in summer, when they are hidden by the leaves. This bud usually contains a complete branch system (a shoot system) and sometimes an inflorescence as well (or alone). The axillary bud will be activated only if the overall hormonal balance of the shoot is changed or disturbed. If the terminal region, the shoot apex (or terminal bud), is lost or destroyed, the buds lower down the stem will become active. For thousands of years the gardeners of Japan have used this method to control the shaping of trees and shrubs, though only in more recent times have botanists begun to learn what really is occurring when a shoot apex is removed from a branch.

Simple Leaves and Compound Leaves

When the leaf is all of one piece, we say it is *simple* or *entire*. An entire leaf may have veins that branch out horizontally to each other from the main vein or *midrib*, in which case the entire leaf is said to be *pinnately veined*. (Such a leaf is seen on the weed on p. 190.) Or the major veins may arise from one point at the base of the leaf where it meets the petiole, and then radiate like the fingers on a hand into the blade. Such a leaf is said to be *palmately veined*. (The leaf of the common household geranium has such a venation.)

The Compound Leaf

The preceding description dealt with a leaf composed of *one unit*—a simple, entire leaf. Not all leaves are so composed. To understand the structure of compound leaves, keep clearly in mind that to be a true leaf—by botanical definition—the leaf must have a bud in its axil.

Now, imagine a large, entire, pinnately-veined leaf. Then begin to imagine that the blade tissue around each of the horizontal veins separates from its sisters until finally each horizontal vein is surrounded by blade tissue, but entirely separate from all others like itself, though all are attached to the midrib. You have created the image of a giant feather, or a *pinnately-compound* leaf (*pinna*, as in pinnate, means feather). The axillary bud is found only at the base of the petiole; none will be present at the base of the newly created *leaflets*. The leaf of species of the genus *Rhus* (sumacs, p. 39) have pinnately-compound leaves.

You may imagine, using the same rules for leaflets as you used for the whole leaf, that a pinnately-compound leaf could be compounded several times; if twice, the leaf is *bipinnate*. Soon such a leaf is very plumelike.

Let's, once more, play divine. Begin this act of creation with an entire, simple, but palmately-veined leaf. Now imagine that the tissue about each fingerlike vein separates from its sister veins as in the pinnately-compound leaf, leaving the new leaf composed of fingerlike leaflets radiating from the most distal portion of the petiole. This is a *palmately-compound* leaf. The relatively well known horse chestnut tree [*Aesculus hippocastanum*] has palmately-compound leaves. The axillary bud of these leaves is found only at the base of the petiole, never at the base of a leaflet.

Other Kinds of Leaf Differences

Even more variation occurs. The margins or edges of leaves or leaflets show a myriad of shape variations. Here are but a few: *undulate* (wavy), *serrate* (with teeth pointed forward), *dentate* (with large teeth).

The whole leaf may also vary in shape from species to species—and sometimes on the same plant! Leaves may have an overall shape that is *orbicular* (round), *ovate* (egg-shaped), *lanceolate* (lance-shaped), *reniform* (kidney-shaped or bean-shaped), *sagittate* (arrowhead-shaped), *deltoid* (delta-shaped or triangular), *spatulate* (shaped like a spoon)—and those are but a few. All of these differences may be of great use to the botanist in making scientific classifications of plants in erecting keys for their identification.

Roots

I have always felt somewhat sorry for roots. Unseen, they are often ignored. Yet no plant could do without them, except for the few specifically evolved to go rootless. Most of our weed species are provided with very healthy root systems—indeed, they apparently have more actively growing roots than most of the more desirable garden species. Among the plants there are two major types of root systems produced: the *taproot system* and the *fibrous root* system. The carrot best typifies the taproot system in which there is one main, conspicuous, dominating root and numerous, sometimes almost hairlike, secondary roots. Select a fresh carrot and study it closely. Note that the bulk of the carrot is one unit, the taproot, and that small, thin side branches seem to come out from wrinkles in the main root. These are the secondary roots. The carrot of commerce [*Daucus carota*] is the same species as Queen Anne's lace (p. 207). Dandelion [*Taraxacum officinale*, p. 135] also has a taproot.

The fibrous root system is typified by that seen on grasses. Here one root does not dominate the system; rather, all the roots are equal, and there are a great many of them. Gingerly pull up some grass (preferably not from your own lawn), and you will find the fibrous root system. You will also find that in pulling up one plant, you usually pull up a region of plants because the roots of fibrous-rooted plants are interwoven. Pulling up tap-rooted plants may not be easy either, and you may leave a small piece way down deep that, in time, will reproduce the plant.

Other Forms of Roots

Tropical plants of wet forests (none of our weeds are in this category) have *aerial* roots and absorb water directly from the moist tropical air. Roots of parasitic plants such as the dodder [*Cuscuta gronovii*, p. 173] are, soon after germination of the seed, sent down into the tissues of the host plant and act thereafter as absorbing agents, taking up water and minerals. Some parasitic plants are green and can, therefore, make their own food; they depend on the host plant for water and minerals only. Dodder not only takes water and minerals but also manufactured carbohydrates because it lacks chlorophyll. One green parasitic angiosperm with which you are familiar is mistletoe, formerly a fertility symbol (and still, apparently, partially one at Christmas).

The Flower

The flower is the basic organ upon which classification of the angiosperms is based. Structurally and morphologically, the flower is a *modified branch system*. The portion of the stem on which the flower is borne is called the *receptacle* and is, fundamentally, a compressed region of the stem bearing several crowded nodes and very short internodes.

The lowest organs of the more typical flower are the *sepals*, which collectively are known as the *calyx*. Usually there is one whorl of them. Just above the calyx on the receptacle is the whorl of floral organs known collectively as the *corolla*, each unit of which is a *petal*. (Sepals and petals considered together are known as *perianth parts*.)

Higher up are one or two whorls of *stamens*, each of which is composed of a thinnish *filament* capped by a four-chambered *anther sac*. It is within the latter that pollen is formed. When the sac is mature and the pollen ready to be shed, the sac opens (dehisces) and sheds the pollen grains. (Don't confuse these grains with the fruit, which is called the *grain*. There is really no similarity.) Pollen may be dry or sticky; the dry pollen usually is carried away by the wind, and the sticky awaits an animal pollinator and then sticks to some part of the animal. It is then carried to the next flower and left. Herein lies the reason that goldenrod [*Solidago* species, p. 145] is *not* the culprit that causes hayfever. This disease obviously can only be caused by a wind-pollinated species of plant. Pollen grains, wind or animal-carried, contain the fertilization (male) nuclei.

The topmost floral organ is the carpel, which when single is called a *pistil*. The pistil is composed of a basal *ovary* and a thinnish *style*, topped by the *stigma*. Within the ovary chamber are produced one or more *ovules*. Later the ovary will mature into the *fruit* and the ovules into *seeds*, but only after fertilization has occurred. Pollen grains land (or are deposited on) the stigma.

Is It a Monocot or Dicot?

A rule of thumb for separating plants into monocots and dicots is by observation of the flowers they produce. Count first the perianth parts (sepals and petals); if you can divide evenly by three, then the flower is probably from one of the monocot species. Usually there will be three or six stamens and often three pistils; though these may be fused, the lines of fusion will still be rather clear on the surface of the pistil. Monocotyledonous species produce flowers whose sepals and petals look alike and are very similar in color and texture. This is not so with species of dicot flowers.

To determine if the flower is a dicot, count the perianth parts. Here, since sepals and petals are usually so totally different in appearance— sepals are generally green, petals are some other color, and each has a very different texture—they can be counted separately. If there are five or if the number of sepals and/or petals and other organs is divisible by four or five, then the plant is probably a dicot. Remember, the leaves of monocots and dicots are differently shaped and veined.

The angiosperms underwent a relatively rapid and various evolution, the exact nature of which is still a question. One rule of thumb is that if the plant has numerous flower parts, i.e., many sepals, petals, stamens, and pistils (as in *Magnolia*), it is a primitive one, while if the plant has a flower with fewer parts, it is more highly (that is, recently) evolved. The numerous parts in primitive flowers are also separate and usually spirally arranged. In more highly evolved species of angiosperms the floral organs are whorled (in circles) rather than in spirals, and the numerous parts have given way to a reduction in numbers of parts—in dicots to fives and fours and in monocots to threes. Even more highly evolved are those plants with flowers that have their five (or four) petals fused together to form a cup or tube or trumpet-shaped structure. The petunia has such a corolla.

Pollination of Flowers

The two major pollinating agents of the flowering plants are insects and wind, but there are others. Water effects pollination in species that grow underwater—for example, *Anacharis (Elodea)*, the waterweed—and many animals other than insects are involved in pollination—among them slugs, birds, and even monkeys! Of course, man has been consciously pollinating plants since ancient peoples carried the male flowers of the date palm to the female flowers as part of what was very likely a religious ceremony. We still consciously pollinate in our huge genetical programs in order to improve agricultural plants. These programs have been quite successful in creating new varieties and in improving old ones.

Natural animal pollination has resulted in flowers that are larger, more showy, and conspicuous, that are more likely to be fragrant and to have rewards for the visiting animal, such as nectar. Indeed, when we say "flower," we tend to think of big, colorful, attractive corollas. The word "attractive" itself gives the story away; they are attractive to us aesthetically, but to the pollinator in far more practical ways.

Wind pollination, known technically as anemophily (literally, wind-loving), has resulted in quite the opposite kind of blossom. It is not at all necessary to attract the wind, so wind-pollinated flowers do not put on a big show. They are usually small, inconspicuous, nonfragrant, contain no rewards, and have even lost their sepals and petals, which have been selected against during their evolution. Grasses are wind-pollinated.

Pollination and Fertilization

When a pollen grain lands on the stigma of the correct species, it will germinate. A tube grows out of the grain and the fertilization nuclei enter the tube and begin their trip. The tube continues to grow through the stigma, digesting its way through the tissues, down the style, and into the chamber of the ovary. It heads for an ovule, enters it, and deposits the nuclei of fertilization within the egg sac of the ovule where the egg nucleus awaits. After fertilization has been accomplished, a new embryo plant forms by cell division within the tissues of the ovule.

The ovule, once fertilization has occurred, itself undergoes growth and slowly, while the embryo is developing within, becomes a seed. At maturity, the seed contains food tissue that surrounds the embryo plant (or is within its cotyledons).

Not only has the ovule been developing after fertilization, but so has the ovary. The latter structure gradually matures into a *fruit*. The mature fruit may be fleshy or dry. If fleshy, it may contain one or more seeds; a fleshy one-seeded fruit is generally called a *drupe*. The cherry and the peach are examples of drupaceous fruits. If the fleshy fruit contains more than one seed, it is a *berry*. Tomatoes and grapes are berries, and special forms of berries are apples, oranges, and fruits like the cucumber. It is interesting to think that the huge watermelon (fleshy and many-seeded) is a ripened ovary grown to great size.

If a mature fruit is dry, there are two possibilities. It may open or it may not. If it opens, then it may open (dehisce) along one line (suture) and is called a *follicle* (the milkweeds, p. 233, produce follicles), or along two lines and is called a *legume* (the black locust, *Robinia*, p. 35, is a legume producer), or along many lines, in which case it is called a *capsule* (evening primrose, p. 143, produces capsules).

However, not all dry fruits open and those that do not fall into two major categories: fruits whose single seed is fused to the ovary wall at but one point and fruits whose seed is fused entirely to the ovary wall. If the seed is fused at but one point, the fruit is an *achene*. Composites generally produce achenes. The true fruits of the strawberry are achenes and the straw*berry* is by no means a berry. The edible portion of the strawberry is the swollen, juicy, and sugary *receptacle*. The tiny, dark dots on the strawberry are the true fruits, the achenes. To demonstrate to yourself that the seed is attached at but one point in an achene, take a sunflower "seed" (an achene) and longitudinally open it with a razor blade. You will see that the one seed within is attached at only one point.

The second type of dry fruit does not open at maturity. Its contents— one seed—are completely fused to the surrounding ovary wall. An example of this type of fruit, the *caryopsis* or *grain*, is the fruit of the members of the grass family. If you dig a kernel (caryopsis) from an ear of corn (inflorescence—each kernel is the ripened ovary of the pistil of one flower on the ear; the corn silk is a mass of styles and stigmata of these ovaries) and hold it with the point of attachment to the ear down and the white shield-shaped area facing you, you will recognize the major portions of the fruit relatively easily.

The tough outside of this grain is the ripened ovary wall, the white shield-shaped area the single cotyledon that is attached to the embryo corn plant (not visible to your eye at this point). You can make this structure out because it appears as a raised line at the center of the cotyledon. The yellowish material around the white cotyledon and embryo is the food material of the grain.

With a sharp razor blade, try sectioning longitudinally through the grain, passing straight down the embryo. Separate the two equal parts of the grain and look inside. For this you may need a low-power magnifying glass. Within, you can see the future leaves and stem of the embryo maize plant (upper portion of the embryo), the attached single cotyledon pressing against the food material. During germination, this structure releases enzymes that change the stored starch to sugar and then absorb the sugar, passing it to the embryo. The embryo is in a period of crisis during germination; it needs energy for the increased metabolism occurring at this time, and can get this energy only from the sugar stored within the seed since the young plant is unable to make its own food. Below the attachment of the cotyledon is the future root of the new plantlet.

When looking into a grain, you are also looking into the source of civilization in the western hemisphere; maize is still, as it was when Columbus arrived in 1492, the basic agricultural grain plant of our hemisphere. The structure of this grain is basically similar to that of the grains of many of the noxious grasses of lawn and garden—for example, crabgrass (p. 85).

The Dicotyledonous Seed

While the grain was a fruit we treated as a seed because of its special nature, the dicotyledonous seed of the bean or pea plant is a true seed. Its ovary wall is a pod (legume) that splits open at maturity and permits the ripened ovules (the seeds) to fall out.

Soak some lima beans or kidney beans for about an hour, and when soft, take a bean out of the water and remove the skin on the outside. This skin is called the *seed coat*; it protects the seed until germination occurs. It is easy to soak off in the bean, but in some species of plants the seed coat is so tough that it is almost impossible to get water through it. After removing the seed coat, you will see two objects that join each other rather perfectly, much the way clam shells do. These are the *cotyledons*, and there are two of them, for the bean is a dicot.

Insert a fingernail between the cotyledons and open them out. Note their thickness. In the bean the seed food has been absorbed into the cotyledons (it was outside the single cotyledon of the corn grain). When the two cotyledons are opened up, one usually breaks away from its attachment. It was attached to the small embryo plant, which can be seen near the surface of the bean near the top. It is whitish and has two small leaves that make it look like a little feather; hence its name, *plumule*. The cotyledons are attached just below those leaves, and right below their attachment is the *radicle*, or future root. When germination occurs, the seed leaves (cotyledons), still folded together, will break the soil first, thus protecting the young regular leaves and the shoot apex between them. Once the root is absorbing water and the first regular leaves have turned green and are photosynthesizing, the plant will be on its own and no longer need to call on the food reserves within the cotyledons. These will then shrivel up and fall away from the plant.

A considerable number of dicot seeds do not store food within the cotyledons (as we saw in the maize caryopsis) but have a similar structural arrangement. Two cotyledons, their upper surfaces lying closely pressed together and their lower surfaces pressed against the food, act as absorbing agents during germination. The cotyledons slowly slide out of the seed as germination progresses, absorbing until the last moment and then unfolding and turning green in the light. The seed of *Ricinus communis* (the castor oil plant) is so structured. These seeds contain ricin, which is a very dangerous poison.

How to Identify the Weeds in This Book

Not all manuals of weeds include keys for their identification. Some list the weeds by the families to which they belong. Others include complex keys that are technically difficult or require the learning of a vast number of meanings and shades of meanings for the numerous descriptive adjectives and nouns. For the reader with little or no previous training in botany or plant taxonomy, this can be very difficult or frustrating enough to inhibit further interest—a great shame, indeed, because weeds are very important plants in our civilization and culture and knowing them is pleasurable.

In this volume every effort has been made to keep complex terminology to the barest minimum and the "system" of identification as simple as possible. Structural information about the flowering plants was dealt with in the last chapter; information that should be of great help in understanding the botany of weeds.

It is recommended that frequent use be made of the clear and beautiful drawings. In each plate a humorous line characterization of the Contessa Giovanna, in her large garden hat, indicates the height and general growth habit of each weed.

Each description includes a recommendation for the removal or destruction of the unwanted plant. *Where herbicides are suggested, keep well in mind that these chemicals can be dangerous—to you, to other desirable plants, and to the environment. It is very helpful to read the label on a herbicide before using it.* You may also write to your Cooperative Farm Agent before selecting or buying a herbicide. New herbicides are constantly being placed on the market and others are withdrawn. Therefore, some of the herbicides recommended in this book may, by the time of publication, have been superseded, withdrawn from the market, or proclaimed dangerous by a governmental agency.

Two major categories of weedy plants are dealt with in this book: flowering plants and nonflowering plants. The only two plants included in the second category are the scouring rush or horsetail [*Equisetum arvense*] and the bracken or eagle fern [*Pteridium aquilinum*]. All the remaining weeds are members of the angiosperms. These are divided into the following general, easy-to-grasp categories: trees, shrubby trees, shrubs, vines, herbaceous plants.

The herbaceous plants are subdivided into monocotyledonous weeds and dicotyledonous weeds. Both subdivisions include only annual species (live for one year and die totally) and those species whose aerial portions die with the onset of winter but whose rhizomes or corms or bulbs continue to live underground.

The monocot weed species are further divided into the broad, general groups of grasslike plants, true grasses, and other monocots; while the dicot weed species are categorized according to flower color in this order: green, yellow-green, yellow, orange, orange-yellow, white, pink or red, blue, and purple. It is recommended that you study carefully the drawings of each of the weeds with any given flower color you are checking and compare with the living plant.

The following is an outline of the contents covered in this book:

DESCRIPTION	EXAMPLE

FLOWERING PLANTS
Trees

1. *Entire*, shiny, toothed leaves; bark (cork) somewhat shiny with horizontal lenticels. — wild black cherry [*Prunus serotina*, **p.** 33]

2. (a) *Pinnately-Compound Leaves*: Each of elliptical, smooth-margined leaflets; pods (legumes) later in season; racemes of white papilonaceous flowers; armed with spines (here modified stipules). — black locust [*Robinia pseudoacacia*, **p.** 35]

   (b) *Pinnately-Compound Leaves*: With toothed leaflets with flared bases; twigs olive green and very stout; large leaf scars. — tree-that-grows-in-Brooklyn or the tree of heaven [*Ailanthus altissima*, **p.** 37]

Shrubby Trees

1. (a) *Smooth Bark (Cork)*: Pinnately-compound leaves with toothed leaflets. — smooth sumac [*Rhus glabra*, **p.** 39]

   (b) *Velvety Dark Bark (Cork with Hair)*: Otherwise similar to 1(a). — staghorn sumac [*Rhus typhina*, **p.** 41]

Shrubs

1. (a) *Smooth, Entire Leaves*: One-half inch long; armed with spines (modified leaves); greenish-yellow sweet-smelling flowers in spring; red berries in fall. — Japanese barberry [*Berberis thunberghii*, **p.** 43]

   (b) *Toothed, Entire Leaves*: One to two inches long; taller shrub when mature; larger spines than 1(a); alternate host in the life cycle of the fungus-induced wheat rust. — common barberry [*Berberis vulgaris*, **p.** 44]

2. *Palmately-Compound Leaves*: Six inches wide; stout prickle-armed canes bending to ground; blackberries in fall. — wild blackberry [*Rubus alleghseniensis*, **p.** 47]

Vines

1. *Entire Leaves*

   (a) *Opposite*: Entire leaves, pinnately veined; sweet-smelling, white, bilaterally symmetrical flowers; purplish-black berries in fall; often climbing on lower branches of trees, shrubs, or infesting in all directions. — Japanese honeysuckle [*Lonicera japonica*, **p.** 49]

   (b) i. *Alternate*: Entire, spade-shaped leaves about one-half inch long; bell-shaped white to pink flowers; climbs by twining. — field bindweed [*Convolvulus arvense*, **p.** 51]

   ii. As in i, but larger—leaves one and one-half to three inches long. — wild morning glory [*Convolvulus sepium*, **p.** 51]

   iii. *Ovate*: Five-veined (parallel); tendrils (two) from base of petiole; armed; green stems. — catbrier [*Smilax glauca*, **p.** 53]*

* Another *Smilax* has very round leaves but is otherwise the same as iii: greenbrier [*Smilax rotundifolia*, p. 53].

FLOWERING PLANTS (Cont.)

    iv. *Three-lobed*: Largish roundish leaves with dentate margins; tendrils (modified branchlets); later in season bunches of blue grapes (berries).

fox grape
[*Vitis labrusca*, **p.** 55]

    v. *Five-lobed*: Large leaves, almost star-shaped; three-forked tendrils opposite each leaf; fruit golfball-sized, light green, and heavily armed.

wild cucumber
[*Echinocystis lobata*, **p.** 59]

2. *Pinnately-Compound Leaves*

    (a) *Three Leaflets*: Shiny, white berries; rootlets attach to support; POISONOUS!

poison ivy
[*Rhus radicans*, **p.** 61]

    (b) *Many Leaflets*: Tendrils at tips of leaves; purple-blue typically papilionaceous flowers; often spreading on ground or hillsides or atop lower plants.

blue vetch
[*Vicia cracca*, **p.** 63]

3. *Palmately-Compound Leaves*

    (a) *Five Leaflets*: Tendrils opposite leaves and branched; dark blue berries in branched small panicles.

Virginia creeper
[*Parthenocissus quinquefolia*, **p.** 65]

HERBACEOUS PLANTS

1. *Monocotyledonous Species*: With generally long, thin leaves, often linear or sword-shaped; parallel venation and clasping bases; stems with scattered vascular bundles; flowers with parts in multiples of three; little or no woody growth.

    (a) *Grasslike Plants*: Leaves not in two rows; flowers without two scales; fruit not a grain.

      i. *In Wet or Marshy Ground*

        (1) Tall plants, four or more feet; long sword-shaped leaves; long reedlike culm bears tightly packed male and female spikes at top; male spike above, female below.

          [a] Male and female spikes touching on culm.

wide-leaved cattail
[*Typha latifolia*, **p.** 69]

          [b] Male and female spikes separated on culm.

narrow-leaved cattail
[*Typha angustifolia*, **p.** 69]

        (2) Short plant, with triangular stems; leaves in three rows; spikes on umbel-like inflorescence.

false nutsedge
[*Cyperus strigosus*, **p.** 71]

      ii. *Dry Ground*

        (1) On lawns, with awl-shaped, cylindrical leaves; flowering stalks end in umbels composed of flowers and bulbils; crushed leaves smell of onion.

wild onion
[*Allium vineale*, **p.** 73]

        (2) Along well-walked pathways; leaves not cylindrical; stems wiry; inflorescence not an umbel; no odor when crushed.

path rush
[*Juncus tenuis*, **p.** 75]

(b) *Grasses*: Leaves in two rows, usually linear with a basal or flattened sheath surrounding stem; stem round; flower a much reduced floret with two scales; separate male and female flowers; fruit a grain (caryopsis).

The members of the grass family (Gramineae) are listed here on the basis of a very broad comparison of the *overall* shape of the inflorescence as it appears *to the eye*:

Slender, bluish-green, fingerlike, tightly packed at end of culm.	timothy [*Phleum pratense*, **p.** 77]
Wider, foxtail-like, yellow-green, more loosely structured than above.	foxtail grass [*Setaria viridis*, **p.** 79]
Flattened spikelets, elliptical in shape and alternating along flowering culm.	perennial ryegrass [*Lolium perenne*, **p.** 81]
Inflorescent culm looser and very feathery in appearance.	little blue stem [*Andropogon virginicus*, p. 83]
Several fingerlike branches (spikes) at top of culm.	crabgrass [*Digitaria sanguinalis*, **p.** 85]
Several fingers, but each spike flatter.	goose grass [*Eleusine indica*, **p.** 87]
Inflorescence opened out in a small panicle (relatively wide leaves for a short grass).	panic grass [*Panicum clandestinum*, **p.** 89]
Inflorescence, a panicle, very opened out.	switch grass [*Panicum virgatum*, **p.** 91]
Large open panicle with slight droop of spikes.	tall redtop [*Tridens flavus*, **p.** 93]
Large less open panicle with drooping spikes on foot-long stems.	brome grass [*Bromus tectorum*, **p.** 95]
Large open panicle not drooping but with many long bristles along each panicular branch.	barnyard grass [*Echinochloa crusgalli*, **p.** 97]
Large open panicle with irregular clusters of densely crowded spikelets at end of panicular branches.	orchard grass [*Dactylis glomerata*, **p.** 99]
Condensed, crowded but large panicle with numerous spikes making panicle look like a fluffy, plumelike structure at end of the very long culm.	reed grass [*Phragmites communis*, **p.** 101]

(c) *Other Monocots*

    i. *Large Plant*: Long, wide, sword-shaped leaves in clusters; flower cluster on a long leafless stalk; flowers lilylike and bright orange.   day lily [*Hemerocallis fulva*, **p.** 103]

    ii. *Retiring Plant*: Short and low; in shade; flowers between two basal-protecting small leaves folded about them; flower with three petals, two very bright blue above and one tiny white petal below.   dayflower [*Commelina communis*, **p.** 105]

2. *Dicotyledonous Species*: The species of dicots will first be separated from each other on the basis of their flower color. Flower colors, in order of their use, will be green, yellow-green, yellow,

HERBACEOUS PLANTS (Cont.)

orange, white, pink or red, blue, and purple.
Within each of the color-grouped species of weeds
further separations will be made, using general
structural characteristics. *Once you know the color
of the flower, thumb through the species having
that color flower and compare the plant with the
drawing.* Even in taxonomy I've known a picture
to be equal to a thousand words!

Dicotyledonous species, with some exceptions
(when dealing with living things there will *always*
be exceptions), have certain characteristics that
separate them from the monocot species. The leaf
of a dicot plant is not usually sword-shaped and
does not generally have parallel veins; instead,
it is usually in possession of leaves with a reticulate
or netted venation pattern. Most often the leaf
of the dicot has a petiole that the monocot
lacks. The flowers of dicots have floral organs in
multiples of fives or fours and their sepals are
usually green. Again, there are exceptions and
evolution has influenced floral development in the
direction of reduction of parts, which makes immediate recognition of the floral organs somewhat
difficult in some flowers; however, as a good rule
of thumb, fours and fives is useful.

(a) *Green-Flowered Weeds*

    (1) Linear or spatulate leaves in rosettes;
veins at first seem parallel.

        [a] With flowering scape terminating   ribgrass
in a very condensed conical spike.   [*Plantago lanceolata*, **p.** 109]

        [b] With tiny flowers covering the   broad-leaved plantain
long thin spike from bottom to top   [*Plantago major*, **p.** 110]
(whole structure looks like a rat-
tail).

    (2) Rhombic-ovate leaves blue-green in   lamb's-quarters
color, lower surface whitish; spike clus-   [*Chenopodium album*, **p.** 113]
ters in axils of leaves; not aromatic if
crushed.

    (3) Oval to heart-shaped (cordate) toothed   cocklebur
leaves with tendency to three lobing;   [*Xanthium pennsylvanicum*,
leaves appear yellow-green; tough   **p.** 115]
hooked burs in fall.

    (4) Usually very large (foot-long), three-   giant ragweed
pointed (trifid), opposite, toothed   [*Ambrosia trifida*, **p.** 117]
leaves, sometimes leaves are five
pointed; leaves very rough to touch;
plants may be fifteen feet high; green-
flowered racemes (male) near top of
plant.

    (5) Leaves bipinnately parted and feathery   common ragweed
in appearance, smaller than (4); with   [*Ambrosia artemisiifolia*,
similar flowering green spikes from   **p.** 119]
axils of leaves near top of plant; plant
a few feet high sometimes.

(6) Leaves darker on top; pinnately dissected; woolly white below; tall; rank odor like chrysanthemum when leaf crushed.

wormwood
[*Artemisia vulgaris*, **p. 121**]

(b) *Yellow-Green–Flowered Species*

  i. *Flat Rosette of Very Large Leaves*

   (1) Large, wavy-edged, lance-shaped leaves in rosette in first year; in second year, at center of rosette a tall branched stalk of tiny yellow-green flowers in drooping spikes.

curly dock
[*Rumex crispus*, **p. 123**]

  ii. *Short Plant*: Sometimes spreading flat, with many small yellow-green heads; crushed leaves smell like pineapple.

pineapple weed
[*Matricaria matricarioides*, **p. 125**]

  iii. *Erect*: Not spreading; with narrow, alternate linear, sessile leaves; milky juice when leaf is broken (latex).

cypress spurge
[*Euphorbia cyparissias*, **p. 127**]

(c) *Yellow-Flowered Species*

  i. *Low-growing*: Prostrate or quite near ground.

   (1) Prostrate.

    [a] Alternate, wedge-shaped leaves; forms a branching mat; leaves thickened (succulent).

purslane
[*Portulaca oleracea*, **p. 129**]

    [b] Opposite, round leaves on short petioles; forms mat on damp ground.

moneywort
[*Lysimachia nummularia*, **p. 131**]

   (2) With rosette.

    [a] Leaves entire, smooth-margined, narrowly oblong to lance-shaped; hairy, two or three leaves on flowering stalk; small cluster of heads atop stalk.

hawkweed
[*Hieracium pratense*, **p. 133**]

    [b] Leaves pinnately lobed and toothed; flowering stalks ending in single quarter-sized heads of flowers (all ray); later, after pollination, ball of feathery fruits you may pick and blow away.

dandelion
[*Taraxacum officinale*, **p. 135**]

   (3) Low-growing, not prostrate, no rosette.

    [a] Large-toothed, cordate (coltsfoot-shaped) leaf; flowering stalks low, scale; blooms very early in spring; single heads at end of stalk.

coltsfoot
[*Tussilago farfara*, **p. 137**]

  ii. *Erect Plants**: Not very low growing.

   (1) Leaves entire, but not dissected.

    [a] Linear, thin, blue-green leaves; sessile; flowers large and bilaterally symmetrical (snapdragon).

butter-and-eggs
[*Linaria vulgaris*, **p. 139**]

* In these yellow-flowered species, some effort has been made to separate them on the basis of their leaf structure and shape. *It is strongly recommended that you study the excellent drawings and compare the weed whose name you seek with the drawings.*

HERBACEOUS PLANTS (Cont.)

[b] Narrow, with several veins visible; grasslike leaves; plant two to four feet high; branching flat-topped clusters of small heads.

grass-leaved goldenrod
[*Solidago graminifolia*, **p.** 141]

[c] Lance-shaped, hirsute (tough-hairy) leaves; short petioles; large four-petalled flowers; plant two to six feet in height.

evening primrose
[*Oenothera biennis*, **p.** 143]

[d] Leaves narrowly lanceolate; serrate; lower leaves with petioles, upper ones sessile; heads small and on several flowering branches that curve outward and down somewhat.

Canada goldenrod
[*Solidago canadensis*, **p.** 145]

[e] Near sea, growing just where highest tide ends; large, lance-shaped, thick, succulent, rubbery.

seaside goldenrod
[*Solidago sempervirens*, **p.** 147]

[f] Big rosettes of white-green velvety lance-shaped leaves; tall flowering stalk arising second year may be branched; stalk covered with velvety leaves.

mullein
[*Verbascum thapsus*, **p.** 149]

[g] Smaller rosette of oblong (sometimes pinnatifid) hairless leaves; stalk leaves sessile, toothed, and pointed; fruits (capsules) round.

moth mullein
[*Verbascum blattaria*, **p.** 151]

[h] Tall plants (one to six feet high or higher), broad, oval to cordate, three-ribbed leaves; large heads (three to six inches) of disc and ray flowers.

sunflower
[*Helianthus annuus*, **p.** 153]

(2) Leaves entire but dissected or lobed.

[a] Pinnately-lobed (pinnatifid) leaves; terminal lobes round; leaf lobe larger than the others; stalk leaves clasping; flowers with four petals; blooms early spring.

yellow rocket
[*Barbarea vulgaris*, **p.** 155]

[b] Palmately-lobed, three-divided, and deeply dissected leaves; dime-sized, five-petalled flower; inner cup of corolla shiny.

field buttercup
[*Ranunculus acris*, p. 157]

(3) Pinnately-compound leaves.

[a] Small, low plant.

[i] Five leaflets, bright yellow papilionaceous petals; pods in clusters of four or five.

bird's-foot trefoil
[*Lotus corniculatus*, **p.** 159]

[b] *Two or more feet high.*

[i] Three- to five-parted leaves; opposite; near top small clusters of heads; in fall small fruits with two teeth stick to clothing in great numbers.

beggar ticks
[*Bidens frondosa*, **p.** 161]

[ii] Very tall flowering stalk from first year's rosette of huge pinnately compound leaves, deeply lobed; stalk leaves similar but large flat-topped umbels on the ridged, stout stalk.

wild parsnip
[*Pastinaca sativa*, **p.** 163]

[iii] Tall and with leaves pinnately dissected (sometimes three times); leaves feathery; yellow button-shaped heads in clusters at top.

tansy
[*Tanacetum vulgare*, **p.** 165]

(4) Palmately-compound leaves.

[a] Three leaflets (trifoliate)

[i] Plant grows relatively flat on ground; appears to have five leaflets with the two lateral leaflets strongly parted but not separate; margins strongly serrated.

common cinquefoil
[*Potentilla canadensis*, **p.** 167]

[ii] Taller plant with leaves similar to [i].

sulfur cinquefoil
[*Potentilla recta*, **p.** 167]

(d) *Orange-Flowered Species*

i. *One to Two Feet High*

(1) Simple, oblong, sessile leaves with one major vein visible; rough-hairy; flower stalk topped by silver-dollar-sized head of orange ray flower and black-purple disc flowered center; touches of red.

black-eyed Susan
[*Rudbeckia hirta*, p. 169]

ii. *Usually Higher Than Two Feet*

(1) Simple ovate to elliptical leaves; obvious pinnately veined; shrubby plant much branched; flowers with bilateral symmetry and spur at back; long thin capsules "pop" when mature.

jewelweed
[*Impatiens biflora*, **p.** 171]

(e) *White-Flowered Species*

i. Plants not green or photosynthetic; white-orange stringy stems twisted about host plant stems; a parasitic angiosperm.

dodder
[*Cuscuta gronovii*, **p.** 173]

ii. Plants green (photosynthetic)

(1) Growing on land; appressed, close to soil or short plants.

[a] Leaves simple.

[i] Leaves linear, blue-green, alternate; all nodes with sheaths; easily mistaken for grass plants.

knotgrass
[*Polygonum aviculare*, **p.** 175]

[ii] Leaves wedge-shaped and in whorls; no nodal sheaths present.

carpetweed
[*Mollugo verticillata*, **p.** 177]

[iii] Leaves ovate to oblong; plants not appressed to ground; branches covered with rows of hairs; upper leaves larger and sessile.

chickweed
[*Stellaria media*, **p.** 179]

HERBACEOUS PLANTS (Cont.)

[b] Leaves compound; trifoliate.

   [i] Creeping plant

      ((1)) Leaflets toothed; five-pet-   wild strawberry
alled flowers white in   [*Fragaria virginica*, **p.** 181]
color; swollen receptacle
at center of each flower.

   [ii] Not creeping plant but low growing.

      ((1)) Leaflets smooth margined;   white clover
each leaflet a round-   [*Trifolium repens*, **p.** 183]
edged triangle.

(2) Plants erect and of more typical sizes. It is recommended that the plates be checked against the plants themselves.

[a] Simple or nondissected leaves.

   [i] Linear lanceolate, opposite   bladder campion
leaves without the three very   [*Silene cucubalus*, p. 185]
obvious veins (nerves) of
Lychnis alba (see below);
flower bottoms are bulbous and
dark veined; petals appear to
number ten due to deep lobing
(really five).

   [ii] Linear lanceolate, opposite   white campion
leaves with three obvious veins   [*Lychnis alba*, **p.** 187]
(nerves), flower base (sepals)
swollen (but less than *S. cucubalus*).

   [iii] Rhombic, toothed, alternate   deadly nightshade
leaves; petals bend somewhat   [*Solanum nigrum*, p. 189]
backward giving flower a "shoot-
ing star" apearance; green berries later become black.

   [iv] Rhombic, pinnately-lobed leaves   horse nettle
armed with sharp epidermal ex-   [*Solanum carolinense*, **p.** 189]
tension; petioles and stems
similarly armed; flowers similar
to *S. nigrum* but larger.

   [v] Leaves oblong, opposite, and   dogbane
seem to be pointed at both ends;   [*Apocynum cannabinum*,
milky juice if tissue is broken;   p. 191]
fruits two fused follicles (at top
and bottom but not in middle),
pendulous.

   [vi] Leaves oblong, small, pinnately   ox-eye daisy
indented; stalk leaves sessile;   [*Chrysanthemum leucanthe-
stalks terminated by silver-dol-   mum*, **p.** 193]
lar-sized heads (one per stalk);
yellow center, white "petals"
(ray flowers).

   [vii] Leaves large (sometimes one   pokeberry
foot long), alternate and simple,   [*Phytolacca decandra*, p. 195]

entire, wavy margined; with stout petioles; tall plants (four to six feet high); inflorescences opposite leaf at node (when present); racemes of small greenish-white flowers; later green, then dark purple berries.

[viii] Leaves large, alternate, oval but irregularly toothed and lobed; ill scented; stout petioles; large white trumpet-shaped five-parted corolla with fused petals; walnut-sized, heavily armed, four-chambered fruits (capsule).

jimson weed
[*Datura stramonium*, **p.** 197]

[ix] Leaves opposite, lance-shaped; large and fused at their bases around the stem (stem seems to pass through them); compact clusters of small heads of white flowers.

boneset
[*Eupatorium perfoliatum*, **p.** 199]

[x] Leaves lanceolate, pinnately-lobed, or toothed; short plant; branched top with racemes of small four-petalled flowers; seed pods tiny, orbicular, notched at top; each seed pod two-chambered with two seeds.

peppergrass
[*Lepidium virginicum*, **p.** 201]

[xi] Stem leaves halberd-shaped (hastate); sessile on stem; basal leaves lanceolate and show various degrees of pinnate lobing, stellate hairs on leaves; racemes look much like *Lepidium*; fruits tiny, heart-shaped, two-celled, more than two seeds.

shepherd's purse
[*Capsella bursa-pastoris*, **p.** 203]

[xii] Broadly ovate, almost cordate leaves; sheath at each node; stout stem that seems jointed and like bamboo; five sepals, which here take the place of petals; small flowers on pendent, branched, narrow panicles (appear as if racemes); followed by tiny white-green papery fruits hanging in feathery clusters.

Mexican bamboo
[*Polygonum cuspidatum*, **p.** 205]

(3) *Pinnately Dissected Leaves or Compound with Three Leaflets*

[a] Rosette leaves much dissected; stalk leaves less dissected; large flat lacy white inflorescence, a compound umbel with one purple-black flower at center; crushed leaves smell spicy and carroty.

Queen Anne's lace
[*Daucus carota*, **p.** 207]

HERBACEOUS PLANTS (Cont.)

[b] Rosette leaves much dissected and stalk leaves equally dissected, almost feathery; loose flower clusters; flowers in heads; crushed leaves with pungent, acrid smell.

yarrow
[*Achillea millefolium*, **p.** 209]

[c] Compound trifoliate leaves; racemes of white papilionaceous flowers; tall and much branched; sweet-smelling plant.

white sweet clover
[*Melilotus alba*, p. 211]

(f) *Pink- or Red-Flowered Species*

　i. *Entire Leaves*

　　(1) Ovate, entire, opposite leaves with three obvious veins; flower clusters from axis of upper leaves; flowers large, pink-white, five-parted, with petals separate at top of corolla and fused into tube thereunder.

bouncing bet
[*Saponaria officinalis*, **p.** 213]

　　(2) Halberd-shaped (hastate) leaves; tiny flowers, red, on spikes that are branched; spikes also red.

field sorrel
[*Rumex acetosella*, **p.** 215]

　ii. *Compound Leaves*

　　(1) Palmately compound with three leaflets; large head of pink blossoms; head brown when pollination has occurred and individual flowers of head turned down; white V on leaflet often.

red clover
[*Trifolium pratense*, **p.** 217]

(g) *Blue-Flowered Species*

　i. *Creeping*

　　(1) Opposite, shiny dark green leaves; quarter-sized, funnel-form flowers.

small periwinkle
[*Vinca minor*, **p.** 219]

　ii. *Not Creeping, Erect*

　　(1) Very rough, hairy leaves and stalk; basal rosette (first year) hairy; branching flower stalk with individual flowers pink before opening, blue when open. Hairs may puncture flesh!

blue weed
[*Echium vulgare*, **p.** 221]

　　(2) Not hairy; rosette leaves may be compound, but leaves on flowering stalk entire and dandelionlike; silver-dollar-sized light blue heads of flowers; composed only of ray flowers.

cornflower
[*Cichorium intybus*, p. 223]

(h) *Purple-Flowered Species*

　i. *Stout Leaves Heavily Armed*

　　(1) Heavily armed leaves; stalk leaves entire and like leaves of dandelion but often more deeply indented along margins; large heads of many flowers (about the size of a quarter across top).

field thistle
[*Cirsium arvense*, **p.** 225]

　ii. *Leaves Not Armed*

　　(1) On dry ground

　　　[a] Orbicular leaves with crenate margins; creeping stems.

ground ivy
[*Glechoma hederacea*, p. 227]

[b] Oblong-ovate, opposite leaves on the square stems typical of members of the mint family; leaves broader at their bases; short plant; corolla small, two-lipped, and bilaterally symmetrical.	heal-all [*Prunella vulgaris*, **p.** 229]
[c] Large, whorled (three to six in a whorl); tall (three to six feet) stems speckled with dark purple; culms capped by large, flat-topped purple flower clusters (corymbs).	Joe-pye weed [*Eupatorium purpureum*, **p.** 231]
[d] Oblong, leathery; if broken at base of leaf, milky juice (latex) appears; umbels of purple five-parted flowers; large follicles opening along one line of dehiscence when mature; parachuted seeds in abundance within follicle.	milkweed [*Asclepias syriaca*, **p.** 233]
[e] Cordate, very large leaf in large rosette (first year); leaf with crinkled edges; tall flowering stalk, much branched, comes up second year in middle of rosette; purplish heads followed by burs that stick to fur and clothing.	giant burdock [*Arctium lappa*, **p.** 235]
[f] Palmately-compound leaves, three leaflets; leguminaceous light purplish flowers; small pods that break into sections (legumes that section this way are *loments*); each section of mature fruit sticks to fur or clothing.	tick trefoil [*Desmodium canadense*, **p.** 237]
(2) On wet soil—pond edges, riversides	
[a] Leaves lanceolate-oblong, sessile, and with short white hairs; long candles of purple flowers in early fall.	purple loosestrife [*Lythrum salicaria*, **p.** 239]

NONFLOWERING PLANTS

A. Aerial stem, jointed and fluted; leaves whorled and usually overlooked as the true leaves.	horsetail [*Equisetum arvense*, **p.** 243]
B. No aerial stem but an underground rhizome only; leaves large and bipinnately compound.	bracken fern [*Pteridium aquilinum*, **p.** 245]

A

C

X1

E

X3

D

B

X¾

F

X¾

Prunus serotina Ehrh.

COMMON NAME   Wild black cherry.
SOME FACTS   Native; perennial; propagates by seeds in stones or pits.
BLOOMS   May to June.
RANGE   Nova Scotia to Florida, west to the Dakotas.
HABITAT   Woodlands, fence rows, roadsides, waste places.

Wild black cherry, when it attains maturity, is a stately ninety-foot tree with a diameter of about four feet. Its wood is highly prized in cabinet-making, ranking close to walnut, and as a result, few taller trees are found and few attain full maturity. Why include such a lovely tree among the weedy plants?

Well, the fruits (drupes) of this lovely tree are beloved of birds, and they drop the completely indigestible seeds along fence rows, from telephone wires, and in passing over waste places. Thickets of much-branching young *Prunus serotina* shoots spring up, and are neither attractively shaped nor valuable in this condition.

The bark is typical of most members of the genus *Prunus*, to which family belong the plum, apricot, cherry, chokecherries, and almond trees. The bark of the younger branches is reddish-brown, with the horizontal markings (lenticels) common to the genus. Young twigs taste aromatic and bitter.

The leaves are bright green and shiny on their upper surfaces. Their lower sides are lighter green. The leaves are broadly lance-shaped to oblong, taper to a point, and have small in-curving teeth on their margins.

Lovely six-inch racemes of white flowers appear in late May or June. Later green fruits (drupes) are seen, and still later, in August or September, the fruits are black and taste a bit bitter though by no means unpleasant. (Children use the green fruits as "peas" in peashooters. I did.)

P. serotina is the most dangerous of the eastern wild cherries. One hundred grams of fresh leaves contain ten times the minimum amount of prussic acid (hydrogen cyanide) considered dangerous. It used to be thought that only wilted leaves contained such high amounts, but recent research has shown that fresh leaves are very dangerous too; the cyanide has volatilized from dried leaves.

The shoots at one year of age are easily pulled out and leave no underground parts to restore growth of the plant.

Wild black cherry [*Prunus serotina*]. **A.** Branchlet showing alternate toothed leaves. **B.** A single entire, toothed leaf which is pinnately veined. **C.** Margin of leaf showing dentition. **D.** Single raceme of flowers. **E.** Single five-petalled white flower. **F.** Raceme of mature fruits, which are drupelets and shiny blue-black. The power of magnification is denoted by X.

B

X1

E

D

C

G

F

A

X5

X½

Robinia pseudoacacia L.

COMMON NAMES Black locust, locust.
SOME FACTS Native; reproduces by seeds, and roots give rise to new plants.
BLOOMS June.
RANGE West, Pennsylvania to Indiana and Oklahoma; south, to Georgia and Louisiana.
HABITAT Roadsides, waste places, open woods.

In 1887 some orphaned lads were playing in the backyard of the Brooklyn Orphan Asylum while fence posts of black locust logs were being erected. Why thirty-two boys should take it into their minds to begin to chew the inner bark of the posts (let us hope not out of hunger!) we'll have to wonder, but chew they did and became badly poisoned as a result. All fell into a stupor, and all recovered.

Powdered bark has poisoned a horse when in an experiment an aqueous extract representing only 0.1 percent of the horse's weight was administered. Evidence suggests that powdered black locust bark contains a heat-labile phytotoxin, and a glycoside has also been found. Fatalities from bark chewing are rare.

The black locust may, under fair conditions, grow seventy-five feet high. Its trunk is straight, long, slender, and coarse-barked. The young branches are armed with tough spines that are modified stipules. These may persist for several years.

Alternate, pinnately-compound leaves are found on the stems and each leaf is composed of seven to nineteen oval to elliptical leaflets, each of which is one to two inches long. These leaves look like those of the acacia; hence the *pseudoacacia* specific epithet.

In June the white flowers make their appearance, occurring in drooping racemes some four to eight inches long. The blossoms are very fragrant and are typical leguminaceous flowers, each with its standard, wings, and keel. These fragrant blossoms, mixed with batter and deep fried, make an excellent base for fritters, and an infusion of them makes a pleasant drink.

In 1751 Peter Kalm, the great Karl Linnaeus' student, then in New York City, reported that he found black locust, apparently planted by citizens, on the streets. However, *Robinia* spreads rapidly and may become a pest that has to be removed. Younger plants may be destroyed with application of Tordon (4 amino-3,5,6 trichloropicolinic acid). Girdling a young tree will also kill it.

Black locust [*Robinia pseudoacacia*]. **A.** Whole branchlet with its alternate (**B**) pinnately-compound leaves. **C.** One leaflet of the pinnately-compound leaf. **D.** A raceme of the typically leguminaceous flowers. **E.** A single flower, its petals removed to show the fusion of nine stamens, leaving one free and the hair pistil at the flower's center; sepals are visible too. **F.** Raceme of mature fruits (legumes) with one bean shown below (magnified five times). **G.** A single spine, which in *Robinia* is a modified stipule.

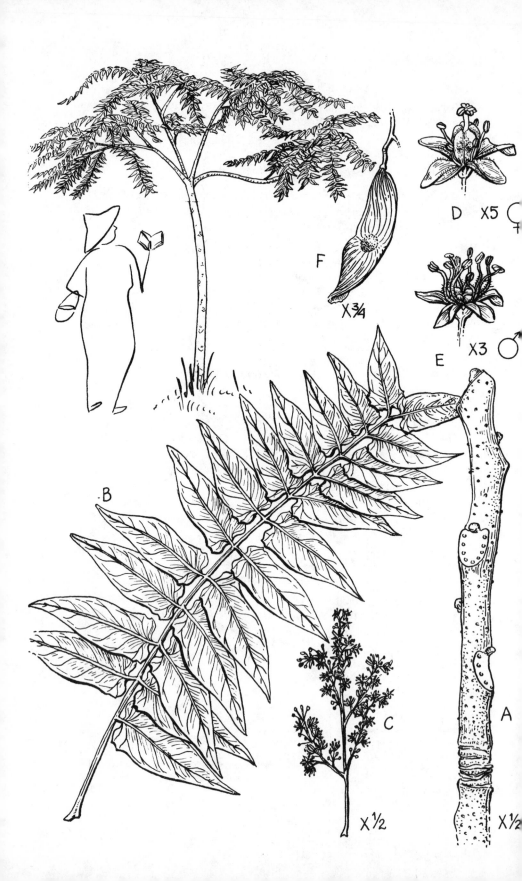

D X5 ♀

F X¾

E X3 ♂

B

C X½

A X½

Ailanthus altissima (Will.) Swingle

COMMON NAMES Tree-of-heaven, stinkweed, tree-that-grows-in-Brooklyn, the weed tree (my name for it).
SOME FACTS Introduced from north China; perennial; reproduces by seeds.
BLOOMS June to July.
RANGE North temperate, oriental.
HABITAT Vacant lots, roadsides, open woodlands.

Though *Ailanthus* is in very ill repute, it deserves credit for one thing—it survives handsomely under the most adverse conditions, viz., in the slums of New York City. *Ailanthus* grows from the very cracks of the tenement walls, providing a touch of green for people who rarely see any. I have seen several trees growing merrily in a garbage- and bottle-filled lot between two grimy walkups. Under these conditions the common name tree-of-heaven seems a bit out of place, but dare one say it has still another common name, the-tree-that-grows-in-Brooklyn? It is said that the laws of Washington, D.C., once forbade planting *Ailanthus* within the District's confines.

The tree arrived in England from north China, where it is native, in 1751, and was introduced to America in 1784, where it has been doing fabulously ever since.

The weed tree grows very rapidly, accomplishing an increase of several feet in one year. Large pinnately-compound leaves, composed of eleven to forty leaflets, are alternately arranged on the stout stems. When they fall, they leave large horseshoe-shaped scars on the olive-tan stems. The leaves are often confused by beginners with those of the smooth sumac and staghorn sumac (p. 41). Each leaflet of a leaf of *Ailanthus* is narrow-oblong and near the base is found a small point lacking in the sumac.

The sexes of *Ailanthus* are separate—trees are either male or female. The females are acceptable in polite society, while the males account for the common name stinkweed. The flowers of the male tree give off a vile, putrid odor, while the panicles of small greenish-white female flowers, after seed set, produce attractive clumps of fruits that are called samaras. Each samara at maturity contains a single seed with wings attached, and these are borne in large clumps, pale gray-yellow in color, not at all unattractive. Each samara is twisted, much like a propellor, and when it falls from the tree, the wind spins and carries it some distance from the tree that bore it.

No special use has been found for this nonpoisonous plant.

To be rid of *Ailanthus*, break the easily snapped twigs until growth ceases.

Tree-that-grows-in-Brooklyn [*Ailanthus altissima*]. **A.** Single stout twig of *Ailanthus*. Note the dotlike lenticels, the huge leaf scars, the axillary buds immediately above the leaf scars, and, near the bottom, the scars left by the bud scales. **B.** The huge pinnately-compound leaves. **C.** Stalk of flowers. **D.** A mature female flower. **E.** The mature staminate (male) flower. **F.** A mature fruit, a samara.

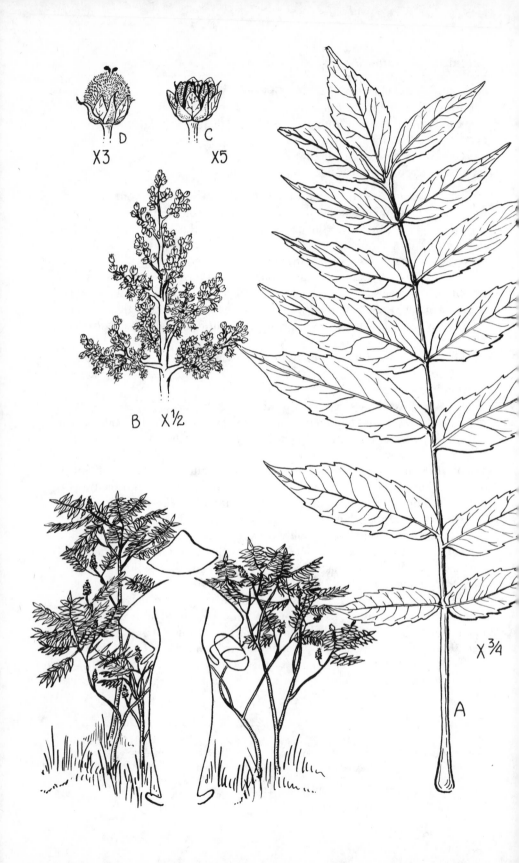

D X3

C X5

B X½

A X¾

Rhus glabra L.

COMMON NAME Smooth sumac.
SOME FACTS Native; perennial; reproduces by seeds.
BLOOMS June to July.
RANGE All of United States; north into British Columbia, south into Mexico.
HABITAT Old fields, roadsides, margins of woods.

Smooth sumac's smoothness would not be marked by the common name or by the specific epithet but for the fact that it has a very hairy cousin, *Rhus typhina*, the staghorn sumac (p. 41). Sumac itself is a corruption of the German *Schuhmacher*, or shoemaker. Shoemakers once used sumac because of its high tannin content.

The genus *Rhus* belongs to the cashew family, to which also belongs a number of valuable and often poisonous plants. The cashew itself, whose delicious nuts are so popular (and expensive), matures in a shell that contains poisons similar in nature to those found in a cousin, *R. radicans* (p. 61). The oil expressed from the shell of cashew is used in making synthetic resins for electrical apparatus, and workers sensitive to the poison in the oil develop symptoms similar to those of poison ivy. *Pistacia* of this family gives us pistachio nuts; mango is also a member of the cashew family, as is the smoke tree of our South. Other species of *Rhus* include poison oak, sumac, ivy, and the tree *verniciflua*, from which the Japanese obtain lacquer. People have been known to react allergically to objects lacquered by the Japanese.

R. glabra, or smooth sumac, is a sparsely branched shrub that may attain a height of eighteen feet. It produces alternate leaves (pinnately compounded) composed of eleven to thirty-one leaflets, each of which is lanceolate or narrowly oblong and two to five inches in length. These leaflets, toothed along their margins, are darker green above than below.

A beverage may be made from the mature red fruits, but because of the shorter hairs on the fruits it is less acid than that made from its close relative, *R. typhina*.

To remove, put on your garden gloves. Defoliate shrub; break main shoot back to ground level. Prevent new shoots from growing.

Smooth sumac [*Rhus glabra*]. **A.** A single pinnately-compound leaf. **B.** A flowering branchlet. **C.** An enlarged single flower. **D.** A single fuzzy fruit. The power of magnification is denoted by X.

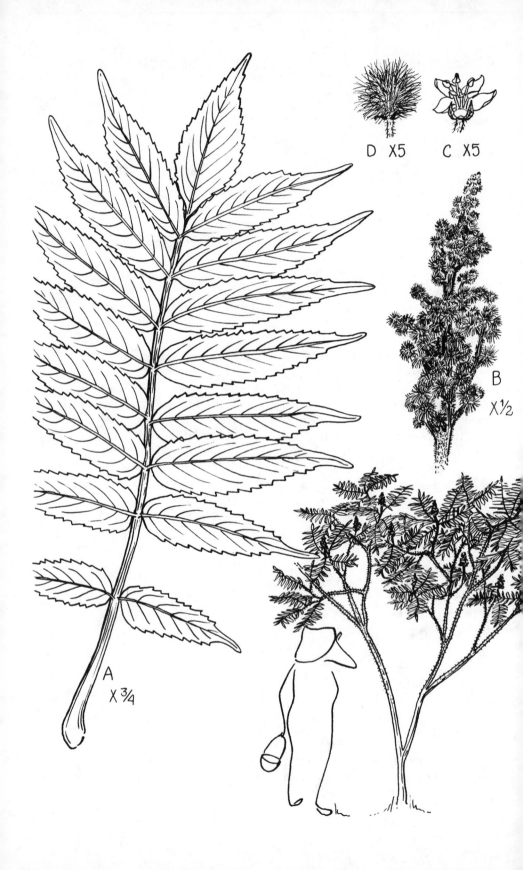

D X5   C X5

B
X½

A
X ¾

Rhus typhina L.

COMMON NAME Staghorn sumac.
SOME FACTS Native; perennial; reproduces by seeds.
BLOOMS June to July.
RANGE Nova Scotia to Quebec, south to Illinois, West Virginia, and the mountains of North Carolina.
HABITAT Dry soil, fence rows, waste places, open lots.

Staghorn sumac usually grows to be a small shrub, but may, under optimal conditions, become a small tree of about thirty feet. The young branches are densely hirsute (hairy), being covered with brown hairs that make the branches appear, after leaf fall, like the new fuzzy horns produced by stags in the spring; hence the name staghorn. The specific epithet refers to *Typha* (p. 69), the fuzzy cattail.

The pinnately-compound leaves have nine to twenty-nine leaflets, each being lanceolate to narrowly oblong and from two to five inches in length, with fine to coarse serrations. The hairiness of the stems extends into the leaf, running up the center of the main leaf rib (rachis).

The erect cluster of unimpressive flowers is followed by more interesting clusters of small, hard red fruits. Their fuzziness is caused by the presence of one- to two-millimeter-long hairs filled with malic acid, the acid of unripe apples. When these are soaked in water, a "pink lemonade" type of beverage can be made that, when sugared, is quite palatable. The American Indians used such a drink as a throat soother.

Peter Kalm, the student of the great eighteenth-century Swedish botanist Linnaeus, says of the "sumach," "The fruits stay on the shrub during the whole winter. But the leaves drop very early in autumn after they are turned reddish like those of the Swedish mountain ash. The branches boiled with the berries yield a black inklike tincture. The boys eat the berries, there being no danger of falling ill after the repast, but they are very sour." Reverend Kalm visited the American Colonies in the late 1740s and wrote a fascinating volume on plant life.

Even more fascinating is John Bartram's information about the use of the red berries of the sumac, *Rhus glabra* (incorrectly called R. *glabrum* by him). He tells us, "The traders informed me that they preserved its perfect blackness [of Indians' hair] and splendor by the use of the red farinaceous or fursy covering of the berries of the common sumach (*Rhus glabrum*). At night they rub this red powder in their hair, as much as it will contain, tying it up close with a handkerchief till morning, when they carefully comb it out and dress their hair with clear bear's oil." Now *there* is a hair dressing to try!

Destroy in the same manner suggested for R. *glabra*. Use gloves.

Staghorn sumac [*Rhus typhina*]. **A.** Single pinnately-compound leaf. **B.** Branchlet of mature fruits. **C.** Single flower. **D.** Single fuzzy fruit.

Berberis thunberghii DC.

COMMON NAME Japanese barberry.
SOME FACTS Introduced from Japan; perennial; reproduces by seeds and creeping roots.
BLOOMS May to June.
RANGE Northeastern states of the United States.
HABITAT Gardens, waste places, borders of woods.

Berberis thunberghii, Japanese barberry, was introduced much later than the European barberry and is a much smaller shrub than its wheat rust–harboring first cousin. Japanese barberry grows to six feet at the tallest, but is more usually seen to be about two or three feet high. It is, for good reasons, a relatively popular hedge plant, used much the way privet is, in the northeastern part of the United States.

It is useful as a hedge plant because of its relatively rapid growth and because it possesses spines. The spines of Japanese barberry are unbranched and about a third as long as those of its taller cousin, but they are long enough and sharp enough to inhibit and instruct for the future that particular type of human nuisance, the hedgecrosser.

The leaves of Japanese barberry are shorter and more delicate in appearance than those of European barberry, being entire and obovate to spatulate, narrowing to a short petiole. Usually there is a purple cast to the leaves of this plant because of the presence of anthocyanin pigments in the sap of the epidermal cells. They may, as some of the leaves of the common barberry, turn bright red in the fall.

Yellowish sweet-scented flowers only eight millimeters wide appear in May or early June and hang in pairs from leaf axils. Later in the season two small berries will take their place in each axil, and still later they will turn red. Because they are relatively dry, unlike those of the taller relative, they are not used in jams and jellies, but they are nevertheless eaten by birds who then proceed to distribute them liberally across the countryside. This accounts for the frequency with which Japanese barberry is encountered far from the confines of a garden.

Berberis (whose scientific name is taken from *berberys*, an Arabic name for the fruit) should be cut at ground level, and because of the armament be sure to wear thick gloves.

Japanese barberry [*Berberis thunberghii*]. A. Branchlet of common barberry [*B. vulgaris*] showing the leaves, which are longer than those seen on B. B. Branchlet of the Japanese barberry. C. The small, yellow, fragrant blossoms; one is shown enlarged (four times). D. The small bright red berries. E. The spines (modified leaves).

Berberis vulgaris L.

COMMON NAMES   Barberry, European barberry.

SOME FACTS   Introduced from Europe; perennial; propagates by seeds and new plants from spreading roots.

BLOOMS   May to June.

RANGE   Northeastern United States as well as north central states.

HABITAT   Waste places, fence rows, gardens; especially fond of gravelly and stony hillside pastures.

The first law passed against a plant in the American Colonies was aimed at *Berberis vulgaris*. In 1754 the Province of Massachusetts passed an act ordering destruction of barberry plants "to prevent damage to English grain arising from barberry bushes in the vicinity of grain fields." Barberry's complicity in the destruction of wheat plants was suspected by European peasants long before the middle of the eighteenth century, and as late as 1847 the botanically important American Dr. William Darlington took a strong position in the frequent arguments that arose over the involvement of barberry in wheat rust. He wrote, "It was formerly a popular belief and one that prevails yet to some extent, that the Barberry possessed the power of blasting grain; the fallacy of this idea has been proved." Thirty years later Darlington's certainty was itself blasted when plant pathologists demonstrated that the common barberry was indeed one of the hosts of *Puccinia graminis*, the wheat rust. Wheat rust passes part of its life cycle within the leaves of barberry, and then goes on to infect the wheat plant. Since 1918 the Department of Agriculture has spent billions in an effort to defeat this member of the club fungi, but total victory has not yet been achieved. In the 70's we are particularly concerned about fungal infections of wheat. Any destruction of our wheat crop or any part of it is instantly reflected in wheat prices, bread shortages and supplies for starving masses, and great heavens, less for Russia!!

Common barberry is a woody shrub that grows four to nine feet tall and branches freely. Its leaves are two to three inches long and are sharply toothed. Each is broadest above its middle and tapers to a short petiole. At a quick glance the leaves do not appear to be alternately distributed along the stem because of the clustering of leaves on short shoots. On the long shoots, alternate distribution is more easily discerned. The under-surfaces of the leaves are gray-green and prominently veined. The stems are heavily armed with three-pronged spines that are modified leaves. Barberry's use as a hedge plant derives from its armed shoots.

Common barberry, because of its greater size, is easily distinguished from its Japanese cousin, *Berberis thunberghii*, but the spines are especially different. On *B. thunberghii*, the Japanese barberry, the spines are not three-pronged but are much smaller (see later). Common barberry also produces long clusters of blossoms, while its more recently introduced Japanese cousin produces short clusters of blossoms (usually two) in the axils of leaves. The common barberry has heavily perfumed six-parted flowers which are small and yellow in color, appearing in drooping racemes composed of ten to twenty flowers each. Red berries appear later in the season and contain from one to a few seeds.

These berries have always been popular for spiced preserves and jellies. Sometimes the berries are cooked in pies and tarts. A syrup may be made from them that, mixed with ice water, will make a good drink for warmer days.

European barberry is relatively resistant to chemicals. You will have to dig it out. Do your best to get the roots or each will give rise to a new plant. Wear thick gloves.

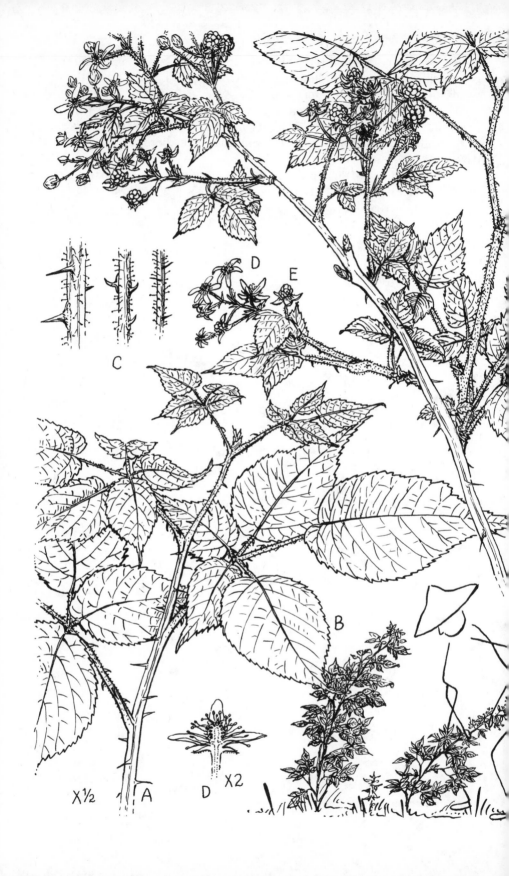

X½   A   D   X2

C

D   E

B

Rubus allegheniensis L.

COMMON NAMES Common blackberry, bramble.
SOME FACTS Native; perennial; reproduces by seeds and rootstocks (runners).
BLOOMS May and June.
RANGE Nova Scotia to Quebec and south to edge of mountains in North Carolina and Tennessee.
HABITAT Open places, edges of woods.

Lovers of the field are familiar with wild blackberry for two reasons: its fruits are delightful and its thorns (prickles) are dreadful. Nature lovers coming upon the plant in fruit will no doubt forgive it its thorns, but I suspect that it is a very unpopular weedy plant during the remainder of the year.

It has long, relatively erect, stout canes covered with very tough prickles. Longer stems arch over from their weight, and where their tips touch the ground, new plants may appear. Bent over, the older canes are relatively tall, coming up to one's knees.

The blackberry is a relative of many plants that produce edible fruits. Among its close relatives are the raspberry, boysenberry, and wineberry. Wineberry's canes are red-purple and absolutely fuzzy with prickles, many of which are not as tough as those of its cousin. The wineberry was introduced from East Asia. The blackberry and its cousins are all members of the great Rose Order.

The blackberry's compound leaves are approximately eight inches long at full growth and are divided into five leaflets, all arising from one point at the far end of the petiole. Such a leaf is said to be palmately compound. Smaller leaves on flowering branches are usually composed of but three leaflets. All leaflets are finely toothed.

The delicious blackberries are preceded by white flowers produced in racemes. Each blossom is about three-quarters of an inch in diameter.

American Indians used the common blackberry as a food and Peter Kalm reports, "They grow about the fields almost as abundantly as thistles in Sweden and have a very agreeable taste."

Because of the armaments, removal by hand is not easy. Fortunately, the plant is sensitive to 2,4,5-T (diluted in oil). Use this on the early summer foliage or on dormant shoots. Even the old living stumps will give way before an attack with 2,4,5-T. If you merely want to thin your bramble patch, dinitro used in a spray will kill that portion of the plant to which it is applied, but will not move to others, nor does it last in the soil.

Common blackberry [*Rubus allegheniensis*]. **A.** Cane, armed. **B.** Large palmately-compound leaves. **C.** The armament, composed of various-sized prickles. **D.** The flower with its five sepals and petals, but more numerous stamens and pistils on a high receptacle. **E.** Young blackberry—not a berry but a cluster of drupelets.

Lonicera japonica Thunb.

COMMON NAME   Japanese honeysuckle.
SOME FACTS   Introduced from Asia; perennial; reproduces by seeds and by creeping stems.
BLOOMS   June and July.
RANGE   Massachusetts to Indiana, south to Florida and Texas.
HABITAT   Fields, thickets, walls, gardens, waste places.

The Japanese honeysuckle, like the English sparrow, was brought to our shores with the best of intentions. At its controlled best, the climbing woody vine is charming and its blossoms lovely to look at and even more wonderful to smell. However, once escaped from cultivation, no amount of fragrance makes up for the outright vileness of *Lonicera japonica*. Large areas of low-growing plants and even saplings will, in short order, become covered by its rapidly spreading stems, which obliterate everything by shading, even climbing and deforming saplings.

This weed travels through the forest in two ways: by runners and by seeds. The latter method is successful because birds like the seeds and scatter them far from the mother plant, but it is the runner that makes this plant truly diabolical in its speed of travel. Runners have averaged twenty-five feet in one year, and it has been reported that one sprout covered more than forty-five feet in one growing season!

The hairy leaves are ovate-oblong and simple, with even margins. Japanese honeysuckle remains green throughout the year and is, therefore, ready for an early start in the spring. Though the leaves lower down the long stems have short petioles, those near the top are sessile and, indeed, sometimes each pair of opposite leaves is fused together (perfoliate).

Yet, the picture is not all bad, for who has never snipped off the base of the lovely sweet-smelling flowers of honeysuckle to suck out the sweet nectar? It is no wonder that people are startled to hear anyone attack this plant, especially if they have been bathed in the fragrance of its blossoms on a quiet night in a beautiful garden.

The plant's yellow or white tubular blooms occur in pairs on short axillary panicles. Each fragrant blossom has five petals fused to form a tube and a two-lipped corolla. Purplish-black berries result from fertilization and each contains two to three small seeds.

Once well established, *Lonicera* (named for the fifteenth-century German botanist, Adam Lonitzer) is very, very difficult to root out. In forests where it has become very well established, bulldozers may have to be called in! In the garden, mow the spreading runners and then burn them (if the law permits). Controlled, careful use of 2,4-D can bring about total elimination of even very extensive infestation.

Japanese honeysuckle [*Lonicera japonica*]. A. The opposite, pinnately-veined leaves. B. The white, fragrant bilaterally symmetrical flowers. C. A sprig showing the fruits, which at maturity are blue-black.

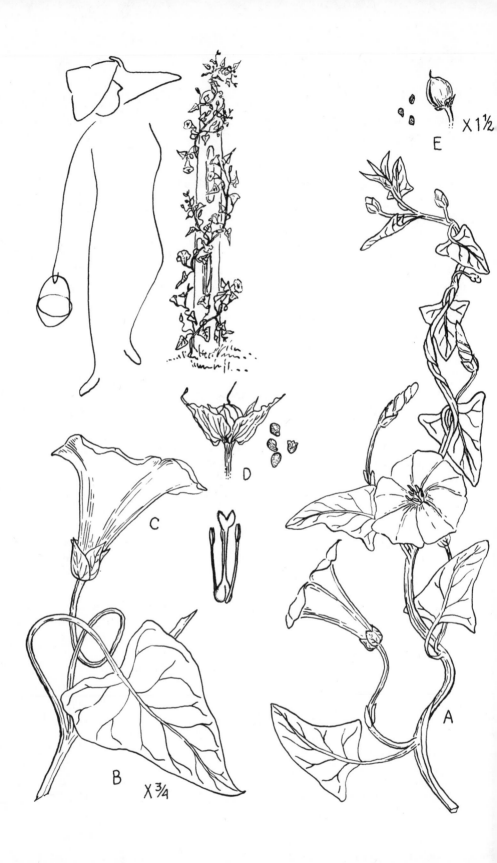

E X 1½

C

D

B X ¾

A

Convolvulus sepium L. (Convolvus arvense L.)

COMMON NAMES Wild morning glory, hedge bindweed, devil's vine, great bindweed, bracted bindweed, hedge lily.

SOME FACTS Native to America; perennial; propagates by seeds and vigorous rootstocks.

BLOOMS June to August.

RANGE Found throughout northeastern and north central United States.

HABITAT Fence rows, meadows, cultivated fields, waste places.

Closely related to the sweet potato, hedge bindweed or wild morning glory produces no gigantic root filled with stored food but is a rather unwanted plant because of its rapid, widespread growth covering over more valuable species. Still, it does produce large bell-shaped flowers.

Devil's vine produces long trailing or twining stems from three to ten feet in length; these may be smooth or hairy. Its rhizomes are not as persistent as those of its first cousin, the smaller creeping jenny or field bindweed, *C. arvense*, with its brittle, cordlike roots whose fragmentation in agricultural fields rapidly spreads the plant.

Smooth leaves, triangular-ovate to halberd-shaped, with divergent basal lobes, are alternately distributed along the twining stems. Their petioles are usually shorter than the leaf itself; on cousin field bindweed the petioles appear longer than the leaves.

Individual largish flowers one to three inches across, pink or white and bell- or trumpet-shaped, arise in the axils of the leaves on peduncles (bases) four to six inches in length. Two heart-shaped bracts are found at the base of the flower hiding the five petals, and later they enfold the four-seeded capsule. Bracts are not seen on *C. arvense*, the field bindweed. Also field bindweed's flowers are smaller, one inch across at best, and several appear in each axil.

For both plants, if chemical treatment is used alone, the underground parts may not be damaged; using chemicals just prior to flowering is best. Constantly cutting off new stems from the roots will eventually starve the underground parts. The plants are moderately sensitive to MCPA salt, 2,4-D amine, and MCPA plus dicambe salts.

Small-flowered morning glory [*Convolvulus arvense*]. **A.** Piece of branch; note its spade-shaped leaves and trumpet-shaped flowers (**c**), which are similar in both species but larger in (**B**), the wild morning glory [*Convolvulus sepium*]. **D.** A single fruit of the wild morning glory with its enlarged protective sepals; a few seeds are seen next to it. **E.** The fruit of the small-flowered morning glory and a few of its seeds.

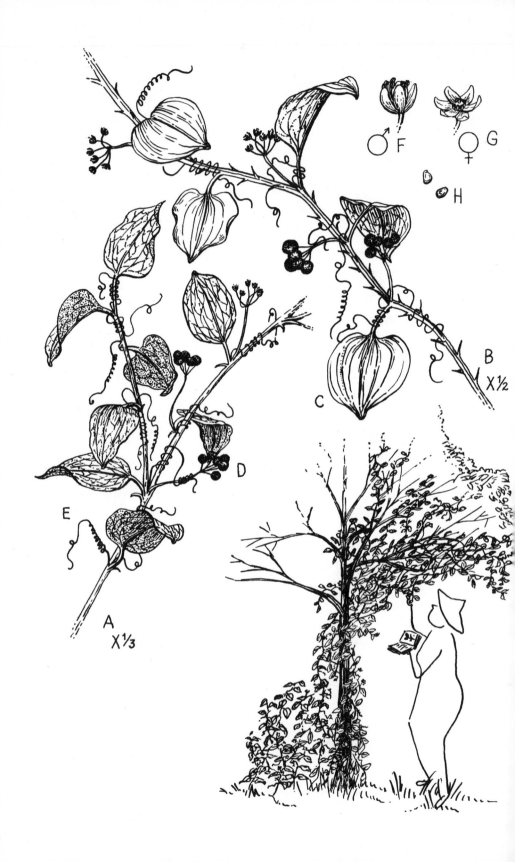

A
X⅓

B
X½

C

D

E

F ♂

G ♀

H

Smilax glauca, Walt. (Smilax rotundifolia)

COMMON NAMES Greenbrier, catbrier, false sarsparilla.
SOME FACTS Native; perennial; reproduces by seeds and tubers.
BLOOMS May to June.
RANGE Common along Atlantic and Gulf coasts.
HABITAT Open woods, thickets, roadsides.

Catbrier is very unpopular with nature lovers, especially those who like to walk through field and forest. Getting into a patch of catbrier can be very disconcerting because the stems are armed and can inflict severe damage on the skin and clothing.

When one first sees any of the greenbriers (there are a number of species), placing it among the monocots does not seem natural. This perennial has round stems with stout curved prickles. Leaves that are broadly ovate sit on petioles, at the base of which is found a persistent tendril that remains on the vine after the leaves have fallen. Another very common *Smilax, S. rotundifolia*, has leaves that are much more round than those of *S. glauca*, plus all the unhappy stem structures of its cousin. Small umbels of six-parted yellow-white flowers emerge from the axis of some of the leaves.

An edible jelly can be made from the roots of *S. rotundifolia*. Dissolved in water, it makes a relatively palatable drink. Bartram reports that the American Indians used these roots after this fashion. True sarsparilla is made from the roots of a tropical *Smilax*. The young shoots of *Smilax* are tasty in a salad or cooked.

Another close relative is *S. herbacea*, which has unarmed stems but armed flowers. These produce a disgusting odor, and for this reason this species has been given the common name carrion flower.

To destroy either *S. glauca* or *S. rotundifolia*, cut the plant down as soon as it appears on your land and treat the roots with caustic soda or carbolic acid.

Greenbrier [*Smilax glauca*] and round-leaf catbrier [*S. rotundifolia*]. **A.** Portion of *S. glauca* stem. **B.** Portion of stem of *S. rotundifolia*. **C.** Single roundish leaf of *S. rotundifolia*. **D.** Cluster of berries. **E.** Tendril. **F.** Small male flower. **G.** Small female flower. **H.** Seeds.

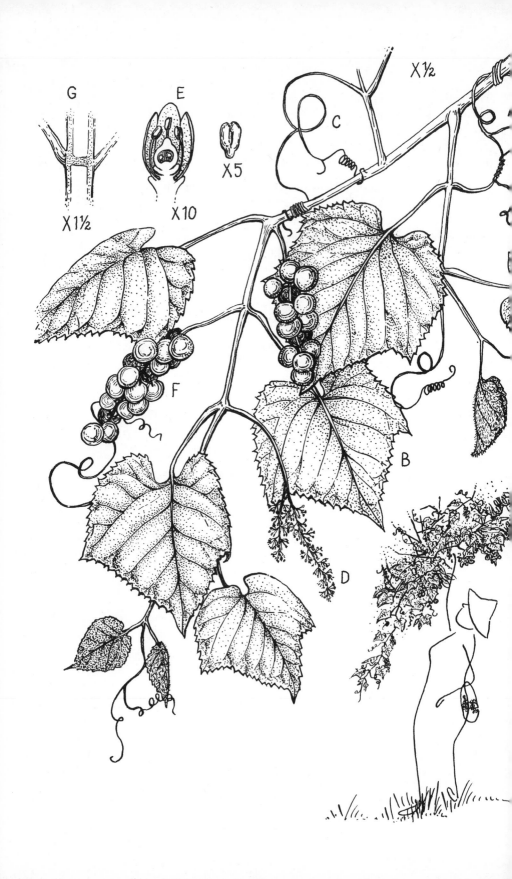

Vitis labrusca L.

COMMON NAMES   Fox grape, wild grape.
SOME FACTS   Native; perennial; reproduces by seeds and rhizome.
BLOOMS   May or June.
RANGE   Maine to southern Michigan, south to South Carolina and Tennessee.
HABITAT   Roadsides, thickets.

Though one occasionally hears that Leif Erikson, the Viking explorer who apparently visited our shores some 492 years before the Genoese Captain Cristoforo Colombo, was deeply impressed by the grape plants he found, the best opinion now agrees he did not find the grape; his *vinber* (or wineberry) was the fruit of the mountain cranberry, *Vaccinium vitis-idaea*, whose fruit is very grapelike. Where those early Vikings made their settlement, in northern Newfoundland, there are no grapes. Where Columbus landed the grape is not a striking part of the vegetation, and he did not include any reference to it in his copious notes on his discoveries in the New World (or as he thought, the Far East).

Some eight hundred years after Erikson, only one year after our own Declaration of Independence, and just before his death, the Philadelphian John Bartram, a prominent colonial citizen with a knowledge of plants, reported from the Indian town of Coolome on the Tallapoosa River in Alabama (about twelve miles east of present-day Montgomery): "During our progress over this vast high forest we crossed extensive open plains, the soil gravelly, producing a few trees or shrubs or undergrowth, which were entangled with grapevines of a peculiar species. The bunches of fruit were very large, as were the grapes that composed them, though yet green and not fully grown, but when ripe rich. The Indians gather great quantities of them, which they prepare for keeping by first sweating them on hurdles over a gentle fire and afterwards drying their bunches in the sun and air, and store them up for provisions."

Our wild grape is parent to many cultivated species now, and let us not forget that the grape is sacred in Western Civilization. From this fruit is made wine, a major beverage-food for thousands of years. So sacred became the fruit of the vine that it was thought to have its own god (Bacchus among the Romans), much the way Ceres was goddess of wheat and grains. Many religions have used wine as a sacred drink, including, of course, the more recently evolved Christianity. Here the juice of the grape is substituted for the blood of the deity during the symbolic service, the Mass.

In 1619 the European cultivated grape was introduced by early colonists

---

Wild grape [*Vitis labrusca*]. **A.** Portion of branch. **B.** The large, entire, but three-lobed leaves. **C.** Tendril, which here is a whole, modified branchlet. **D.** Single group of flowers arising opposite leaf. **E.** An enlargement of floral sections. **F.** A cluster of the mature fruits (berries). **G.** The diaphragmed stems of grape.

with little success. It was not until the nineteenth century that native grapes (e.g., *Vitis labrusca*) came to be domesticated and good characters from V. *vinifera*, the cultivated European species, incorporated. V. *vinifera* is the cultivated grape and all modern grapes, with few exceptions, are apparently direct or hybrid descendants of it. It was the wine grape of ancient southwestern Asia, in cultivation long before the time of Christ and oft-mentioned in the Bible. It was not realized until the nineteenth century that the hardiness of native American grapes could be used to achieve successful grape growing in the United States. V. *labrusca*, wild or fox grape, is one of the several native American species used.

The stem, characterized by its shreddy outer tissues and a diaphragmed pith, lacks much of the supportive, mechanical tissues possessed by the plants on which it so often is found growing. It clutches its support by means of tendrils. Despite the lack of supportive tissue, one shoot may reach one hundred feet in length. While this kind of growth aids grape growers and is delightful for those with arbored gardens, it is not all for the good.

In a special sense *Vitis labrusca* can be a parasitic plant because it covers the foliage of its living support, thus preventing sunlight from reaching the host's leaves. In grabbing the light for itself, it prohibits its living support from making its own food and the support will soon suffer serious growth limitations or die. In this way, dense, impenetrable tangles of grape may not only spread out along the ground, suppressing all other plant life, but also grow up and over the leaf cover of shrubs and lower trees, eventually covering acres of herbs, shrubs and trees. In the wild, whole areas, even portions of forests, may be threatened.

The large, eight-inch leaves are as wide as they are long and are shallowly three-lobed. On their lower surfaces the hairs are so dense that the surface itself is invisible. The serrations along the edges are various sized. These large leaves are opposite and simple. The flowers, borne in raceme-like panicles, are small and relatively fragrant. Each contains stamens which are found exactly opposite their petals. The fruit which follows pollination is popularly called a grape, but is botanically a berry, that is, it is a fleshy fruit which contains more than one seed.

John Gerard, in his herbal printed in 1597, gave much space to the grape vine and wrote, "Almighty God for the comfort of mankinde ordained Wine; but decreed withall, That it should be moderately taken, for so it is wholesome and comfortable: but when measure is turned into excess, it becommeth unwholesome, and a poyson most venemous. Besides, how little credence is to be given to drunkards it is evident; for though they be mighty men, yet it maketh them monsters, and worse than brute beasts. Finally in a word to conclude; this excessive drinking of Wine dishonoreth Noblemen, beggereth the poore, and more have beene destroied by surfeiting therewith, than by the sword." And we all know he jesteth not!

Yet, it is not impossible that, withal, you will want to rid yourself of wild grape. This may be done either by cutting the main stem and checking to be sure new shoots are not coming up, or by using a 1 percent solution of 2,4,5-T in water in the spring when the foliage has just unfolded. Removal of a band of bark one foot above the base of the climbing stem, with application of a 4 percent solution of 2,4,5-T in an organic carrier, should kill the plant.

H X¾

G X¾

E

F

D

B

C

A X½

Echinocystis lobata (Michx.) T & G

COMMON NAMES Wild cucumber, balsam apple, mock apple, four-seeded bur cucumber.

SOME FACTS Native to the United States; annual; reproduces by seeds.

BLOOMS July to September.

RANGE Widespread in northeastern U.S., and westward to Saskatchewan and Texas.

HABITAT Fence rows, damp rich soil, thickets.

Vines climbing unsupervised can act as parasites (see *Vitis*, p. 55, *Lonicera*, p. 49) in that they cover the leaf surfaces of the supporting plant, denying it sunlight, and thus stunt or kill the plant on which they lean. Balsam apple is no exception; it is often found in nature covering the leaf surface of its support.

Under control, however, the wild cucumber vine is an interesting and attractive plant when grown for quick shade in an arbor or to cover a fence or an old tree stump. Its fifteen- to thirty-foot-long stems are smooth except at the nodes, where a few hairs may be seen. Its alternate leaves, generally orbicular in outline, are palmately lobed and veined. There may be three to seven leaf lobes, but five is the most frequent number. A sinus is seen where the relatively long petiole meets the leaf. Opposite the leaf is found a three-forked tendril.

In the axil of the tendril appear the staminate and pistillate flowers. They are white, fragrant, five- or six-parted, with the staminate flowers attracting the attention in prominent racemes, while the pistillate flowers, often solitary, emerge from the same axis but are usually inconspicuous.

The ovoid fruit is very interesting. Approximately two inches at maturity, it pops open at the top and the inch-long seeds are suddenly and forcibly ejected. Beware of the old, inner, fibrous-netted dried fruit. Its spines can jab the unsuspecting painfully.

To keep wild cucumber under some semblance of control, pull out the seeds or snip off the female flowers when they appear. Snipping the main stem will also rid you of this vine.

Wild cucumber [*Echinocystis lobata*]. A. Portion of stem. B. Palmately lobed, five-pointed leaf. c. Three-forked tendrils. D. Axillary panicle of male flowers. E. Single sterile flower. F. Fertile flower. G. Prickly fleshed fruit with two large seeds. Very spiny at maturity when dry.

Rhus radicans L.

COMMON NAMES Poison ivy, markweed.
SOME FACTS Native; perennial; reproduces by seeds.
BLOOMS Late May to July.
RANGE Nova Scotia to British Columbia, south to Florida, Arkansas, and Utah.
HABITAT Roadsides with stone walls, banks and waste places; climbing trees.

Few plants in the United States are more disliked or feared; yet this plant that is despised by all is rarely recognized by any. Many people will tell you exactly what *Rhus radicans* looks like—while standing in the center of a large poison ivy patch!

The leaf is composed of three shiny leaflets (note *leaflets*, not leaves), and the old adage, "Leaflets three, let it be," is one worth heeding. The vine is often confused with the perfectly charming and innocuous Virginia creeper (*Parthenocissus quinquefolia*), which has five leaflets per leaf and should not be destroyed; therefore, it may be wise to remember a new adage: "Leaflets five, let it survive" (p. 65).

All parts of the three-leaved ivy (observe that this common name contains a botanical error) are poisonous except for the pollen grains. Tearing and breaking the stems even in midwinter is dangerous to man and other closely related primates, though animals situated lower on the evolutionary ladder seem to be relatively safe. Birds eat the small white berries without ill effects—to themselves (later they "plant" the seeds far from the original vine and thus spread the plant).

It is said that the juices from jewelweed (*Impatiens biflora*, p. 171) will prevent poisoning by this plant or halt its progress in one who has become contaminated. Indeed, some drugstore preparations contain substances from jewelweed. Taking a bath in water to which crushed, boiled jewelweed leaves have been added is said to give excellent protection.

You may wonder why poison ivy also has the common name markweed. Peter Kalm in 1748 wrote about this plant, "When the stem is cut, it emits a pale brown sap of disagreeable scent. This sap is so strong that the letters and characters made upon linen with it cannot be removed, but grow blacker the more the cloth is washed. Boys commonly marked their names on their linen with this juice."

2,4-D or 2,4,5-T (amitrole or amino triazole) will kill any plants that spring up on your property, and even those long established. In tearing out even dead plants, exercise tremendous care.

Poison ivy [*Rhus radicans*]. A. Shiny, pinnately-compound (three leaflets) leaf B. Cluster of white berries. C. Modified roots, which aid in holding *Rhus radicans* to its support.

Vicia cracca L.

COMMON NAMES   Wild vetch, tufted vetch, blue vetch, cow vetch, bird vetch, cat peas, titters, tine grass.

SOME FACTS   Introduced from Eurasia; perennial; reproduces by seeds and rootstocks.

BLOOMS   June to July.

RANGE   Northeastern United States and eastern Canada; also found on Pacific Coast.

HABITAT   Fields, meadows, waste places, roadsides, especially along banks.

The lovely one-sided axillary racemes of the purple-blue flowers of the blue vetch enliven the banks along the sides of our ever-increasing parkways. Rain leaches nitrates from roadside slopes and the root tubercles on wild vetch, a legume, make nitrates, permitting this pretty weed to flourish. It is also grown to provide a fine forage and good hay, but because of its tough creeping roots, it is difficult to remove from areas where it has become undesirable.

A slender climbing or spreading two- to five-foot stem may also sport appressed hairs, though not always. The tendrils at the tips of its pinnately-compound leaves help tufted vetch to climb—on other plants and smother them—or spread on the ground and shade out grass or other desirable plants under it. Each leaf is composed of eight to twelve pairs of thin, oblong-lanceolate leaflets, each of which ends in an abrupt tiny point, the leaf itself ending in a tendril. The whole plant is a soft olive green in color.

The numerous purple-blue typically leguminaceous flowers on the one-sided axial racemes are approximately as long as a leaflet, i.e., about one-half inch, and hang bent outward on their stalks. The pods (legumes) that appear after fertilization are approximately an inch in length.

The plant is moderately sensitive to MCPA salt and to 2,4-D amine and a number of other weed killers. Several applications may be necessary to rid an area of tine grass.

Blue vetch [*Vicia cracca*]. A. Portion of the vine. B. Pinnately-compound leaf opposite an inflorescence. C. The tendril is at the end of the leaf. D. The flower cluster (raceme). E. One of the papilionaceous flowers. F. After fertilization a legume appears. G. A seed enlarged. The power of magnification is denoted by X.

E
X1

F
X1

D

C

B

A

X½

Parthenocissus quinquefolia (L.) Planch.

COMMON NAMES Virginia creeper, woodbine, American ivy, five-leaved ivy.
SOME FACTS Native; perennial; propagates by seeds.
BLOOMS June.
RANGE Maine to Ontario and south to Florida and Texas.
HABITAT Moist soil, rich woods in disturbed areas, along fencerows and walls.

Virginia creeper is not a weed—not by most definitions. Then why include it in this book? It has been included in an effort to protect this lovely vine so often mistaken for the far less valuable poison ivy. Poor Virginia creeper has been burned, hacked and uprooted, all quite unnecessarily, because most people are unable to tell the difference between this pleasant plant and the more noxious vine. Compare the figures of both; poison ivy is found on p. 61.

Both poison ivy and American ivy have compound leaves, but the true similarity stops there, even in the compoundedness of their leaves. Virginia creeper usually has long-petioled, palmately compound leaves composed of *five leaflets*, while poison ivy's compound leaf is composed of *three leaflets*. Virginia creeper's leaflets are each a dull, darkish green, while those of poison ivy are lighter green (sometimes almost yellow green) and *shiny*. Each leaflet of Virginia creeper's five is elliptical to obovate and serrated beyond its middle.

Later in the season, white berries are found in clusters on poison ivy while Virginia creeper is sporting berries that are nearly black. Each of these berries contains 1 to 4 seeds and was produced by a flower whose parts were in multiples of five. The flowers are produced in panicles that are usually longer than they are wide; the panicles are produced opposite the leaves.

Parthenocissus quinquefolia is not selective of its companions and may be found climbing in the same place or on the same wall or tree as *Rhus radicans*. Both climb with the support of special organs, poison ivy using modified roots and Virginia creeper by means of adhesive disks at the end of much-branched tendrils.

The generic name comes from two Greek words: *parthenos* meaning virgin (as in the Parthenon, a temple devoted to Athene, a virgin it was said), and *kissos* meaning vine. *Quinquefolia* means having five leaves, which this plant does not but five leaflets.

Should you still want to get rid of this vine you will find it quite sensitive to foliar applications of 2,4-D and 2,4,5-T esters applied at two pound acid equivalent for fifty gallons of water. Cutting the main trunk and preventing regrowth ought to kill the vine as well.

Virginia creeper [*Parthenocissus quinquefolia*]. **A.** Portion of vine's stem. **B.** A palmately compound leaf with five leaflets. **C.** Cluster of small blossoms. **D.** Discs attaching vine to wall or other plant. **E.** Single fruit. **F.** Seed.

HERBACEOUS PLANTS

Monocotyledonous Species

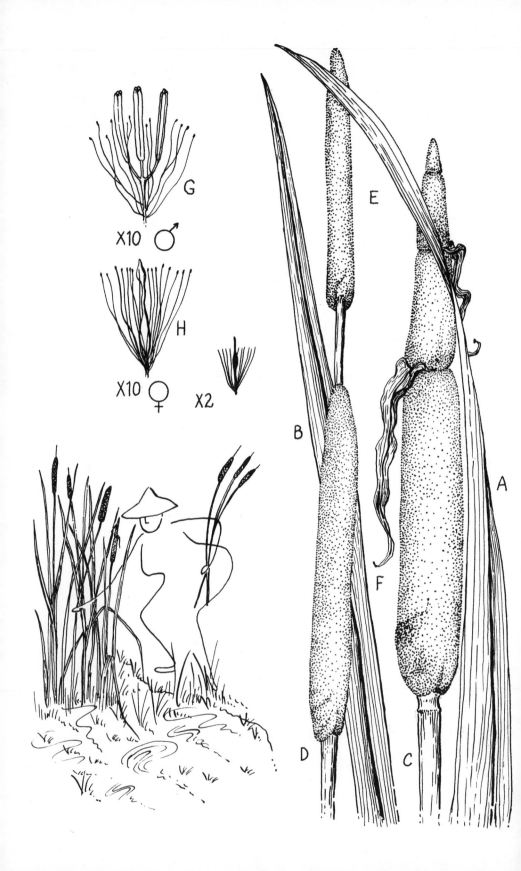

A

B

G

X10 ♂

H

X10 ♀   X2

E

F

C

D

Typha latifolia L. (*Typha angustifolia*)

COMMON NAME   Cattail (bulrush or reedmace in England).
SOME FACTS   Native; perennial; reproduces by seed and rhizomes.
BLOOMS   July and August.
RANGE   Throughout North America (except in extreme north).
HABITAT   Marshes, very wet soil, drainage ditches along roads.

As children, we called the fruiting stalks of cattail "punks," and after drying them in the sun, we lit them and let them burn slowly savoring their odor and pretending we were smoking cigars (which they approximate in size and color). Are there any such innocents around anymore?

Cattail may be wide-leaved or narrow-leaved, the narrow-leaved form being considered a separate species, *Typha angustifolia*. The two species look alike and grow under similar conditions, though the wide-leaved form is larger.

But wide- or narrow-leaved, cattail grows abundantly in marshes and likes to have its roots under water. Stands of cattail composed of many plants are usually seen because of the rapid growth of its rhizome. From these arise one- to nine-foot-long culms with long, half-inch wide, clasping, linearly sword-shaped leaves (narrower on *T. angustifolia*).

The "punk" is a spike of mature fruits (achenes) and their attendant hairs and was, before fertilization, the spike of female flowers. The male spike is borne immediately above the female and these flowers drop from the culm after their pollen has been shed.

The green inflorescent spikes, when cooked with salty water, make a tasty vegetable, and the pollen, when mixed with wheat flour, can be used in making pancakes. Russians who live along the Don love the young shoots, cooking them the way we cook asparagus or eating them raw. The rootstocks contain an edible starch and the leaves are used in making rush seats. Even the fuzz from the mature fruits can be used as a kapok substitute, and a paper is made from the culms.

John Gerard, the sixteenth century herbalist, has this to say about "catstails": "The soft downe stamped with Swines grease well washed, healeth burnes or scalds with fire or water." Perhaps.

A dose of ten to thirty pounds per acre of Dalapon applied one or more times from May to September will generally prevent growth for twelve to eighteen months. The Dalapon should be applied to the leaves. Always use special care when treating a water plant with lethal chemicals because so many living things may be adversely affected.

Wide- and narrow-leaved cattails [*Typha latifolia* and *T. angustifolia*]. A. Wide leaf of *T. latifolia*. B. The narrower leaf of *T. angustifolia*. C. The taller and thicker inflorescence (both male and female) of *T. latifolia*. D. Thinner *T. angustifolia*. E. The higher male flowers, which fall after pollen is shed. F. The female flowers, which remain for a long time on the stalk until the seeds are mature and are carried away by the wind. G. Individual, much reduced, female flower. H. A single enlarged male flower showing its three stamens (a monocot). Note also the parallel veins on the leaves.

X1½

A

B

C

Cyperus strigosus L.

COMMON NAMES Straw-colored cyperus, false nutsedge, lank galingale.
SOME FACTS Native; perennial; propagates by seeds and by tubers; corm-like.
BLOOMS July to September.
RANGE Widespread in eastern United States, but also found locally on Pacifiic Coast.
HABITAT Along streams, in damp meadows, poorly drained fields, ditches.

Sedges are first cousins to the grasses but may be separated from them in a number of ways. First of all, if you study the stem of both, you will find that the stem of a sedge is triangular, while the stem of a grass is round. Snap both stems and you will note that the stem of the sedge is solid throughout its length, while that of the grass is solid only at the nodes and hollow between them.

A comparison of the distribution of leaves along the sedge and the grass stem reveals that those of the sedges are three-ranked and those of the grasses two-ranked. (Three-ranked means that each leaf is one-third the circumference of the stem away from its higher neighbor or its lower neighbor. Two-ranked means each leaf is separated from the others by one-half the circumference of the stem; thus if one leaf is on the "right" side as you look at the stem, the next leaf above or below will emerge from the "left" side.)

Grass leaves are flat, while sedge leaves are grooved or somewhat V-shaped. Relatively few sedges have become weedy (unfortunately, *Cyperus strigosus* is an exception), whereas many grasses are included in this designation.

Arising from a hard cormlike base that is slightly swollen, *C. strigosus* stands two to three feet high and will be found only on poorly drained sections of land. As a member of the sedge family, it is related to the very common hummock sedge of swamp regions and to the bulrushes (*Scipous*). You may know *C. alternifolius*, the umbrella leaf. The ancient Egyptians used the pith from the stems of *C. papyrifera*, which they found along the banks of the Nile, to make the first paper.

The flat spikes of wind-pollinated flowers are borne in umbels, which may be either simple or compound. An involucre of three leaves, all longer than the umbel, surrounds the umbel.

The leaves are flat (not V-shaped, as are the leaves of many members of the sedge family) and smooth, and most of them arise near the base of the stem whose height they match.

Hand pulling should remove the plant if it has become a problem.

False nutsedge [*Cyperus strigosus*]. A. Stem encased in clasping leaves. B. Triangular stem. C. Flattened, straw-colored spikelets.

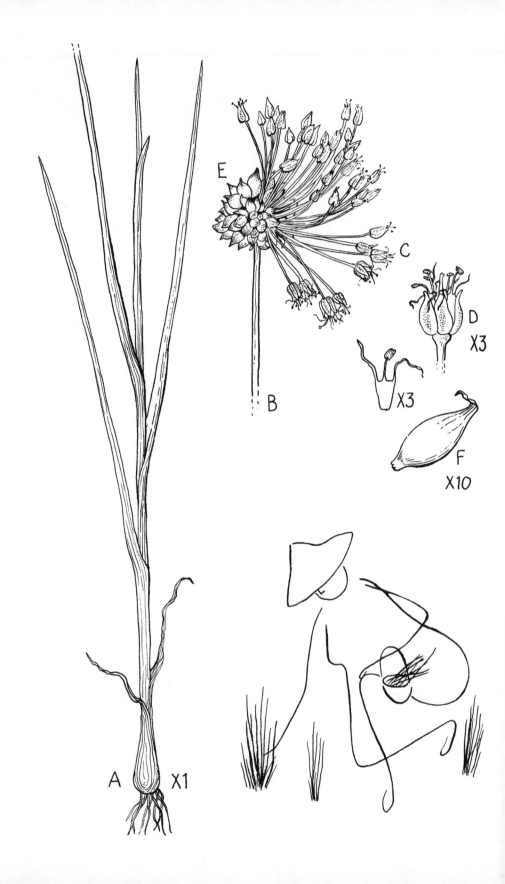

A X1 B C D X3 E F X10

Allium vineale L.

COMMON NAMES Wild garlic, wild onion, field garlic, crow garlic.
SOME FACTS Introduced from Europe; perennial; reproduces by bulbs and
   bulblets formed along flowers, rarely by seeds.
BLOOMS May to June.
RANGE Massachusetts to South Carolina to Mississippi.
HABITAT Fields, meadows, pastures (prefers sandy loam).

Allium vineale, wild garlic, is a close relative of the table onion (*A. cepa*), and like it has a pungent odor that brings tears to the eyes. If you own any milk cows that have grazed on wild garlic (common in pasturelands), this weed may bring additional tears to your eyes because it spoils the cow's milk. The pungent, bitter, tear-inducing constituent of onion is allyl sulphide, and if cows eat members of the genus *Allium*, this constituent will reappear in their milk or in their muscle tissue. As you might expect, cattlemen try to keep wild garlic out of their pastureland.

It might be well to point out here that consumption of a great quantity of onions (*A. cepa*) may result in a temporary but substantial drop in your red blood cell count, and if you are already anemic, this could be a serious health hazard or even lethal.

The slender, pointed, hollow, dark green leaves, round in cross section, emerge from an underground bulb. Wild onion produces two kinds of bulbs: soft bulbs, which start growth during their first fall; and hard bulbs, which are dormant in winter and then germinate during their first spring. Hard bulbs may remain dormant for longer than one year.

From late May to late June the flowering stalk of this member of the Liliales may be seen to arise from the center of the leaf cluster. Small pink-purplish six-parted flowers are produced in umbels. These flowers are replaced by small bulblets, each of which is tipped by a slender filament or tail. One flower seed head may contain thirty to one hundred bulblets, each approximately the size of a wheat grain. Indeed, wheat grain is occasionally adulterated by these bulblets (called cloves by some), and bread made from wheat ground with these bulblets will be unmarketable.

Wild garlic may be removed from the lawn by applying MCPA or 2,4-D amine; however, the plant may show moderate resistance to this treatment. Repeated applications should be used for older plants. Ada Georgia, who wrote a weed book at the turn of the century, recommended applying carbolic acid to the plant with an oil can, a few drops to each plant, and the recommendation is still a good one.

Wild onion [*Allium vineale*]. A. One plant, showing awl-shaped, blue-green leaves and onion swelling at bottom of plant. B. Single cluster of flowers and bulbils. C. An individual flower. D. Flower enlarged. E. Bulbils forming in flower cluster. F. An individual bulbil enlarged.

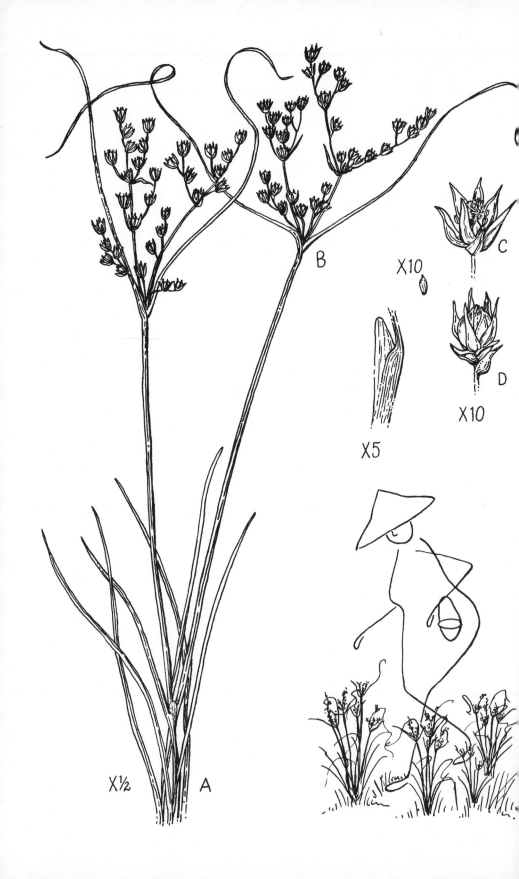

B

C

X10

D

X10

X5

X½   A

Juncus tenuis Willd.

COMMON NAMES   Path rush, wire grass, yard rush, field rush, poverty rush, North American rush (by the British).
SOME FACTS   Native to United States; perennial; reproduces by seeds.
BLOOMS   June to August.
RANGE   Widespread in United States and Canada.
HABITAT   Waste places, along paths in fields, roadsides.

Walk down a dew-laden path early some morning and then check your shoes. Chances are that a sizable number of small orange-brown gelatinous seeds will have stuck to their sides. These are the seeds of the path rush, a plant found abundantly along the paths man makes through fields.

The North American rush, as it is known in Britain, migrated to Europe, appearing first in Belgium in 1825 and then in Britain in 1883. It is now common in 35 percent of the counties in Britain and has also spread to New Zealand, Australia, and Bermuda.

Someone not trained in botany might easily think that *Juncus* is a member of the grass family; however, it is not a grass. Its flowers have sepals and petals, and though both are harsh and strawlike, they are real sepals and petals. Furthermore, the fruit is a capsule filled with small seeds. A proper grass produces a grain (caryopsis) that contains but one seed.

This slender rush seems happy on wet or dry soil. Its thickly tufted stems may be six inches to two feet tall and are very thin, round, and wiry enough to spring back quickly after being trod upon. Its complete flowers appear in irregular clusters between two long, flattened leaves at the top of the stem. Each is composed of three green sepals and three green petals, lanceolate and sharp-pointed.

Path rush is very sensitive to sodium chlorate, and moderately so to boric oxide.

Path rush [*Juncus tenuis*]. **A.** Whole plant, showing the linear leaves. **B.** The stalked inflorescences. **C.** An enlarged drawing of a single flower. Unlike the grasses or sedges that this plant looks like, the flower has all parts. **D.** The capsule mature and ready to fall. The power of magnification is denoted by X.

Phleum pratense L.

COMMON NAMES   Timothy, Herd's grass.
SOME FACTS   Introduced from Europe; perennial; propagates by seeds; most important hay grass in the United States.
BLOOMS   June and July.
RANGE   Temperate regions of both hemispheres.
HABITAT   Humid regions of northeastern United States, open lots; likes clay loams and lightly textured sandy soils.

While timothy may be the most important pasture grass cultivated in the United States, it has escaped from cultivation (especially in the temperate regions) and become a weed by growing where it is not wanted. This tall grass grows well in the cool, humid climes of the Northeast and South into the cotton belt. It is also found as far west as the 100th meridian and in the humid regions around Puget Sound, along the coastal area of the Pacific Northwest, and in the valleys of the Rocky Mountains.

Identifying grasses is not always the easiest of tasks, but once timothy has put out its crowded terminal green spike (a tightly contracted panicle), it is immediately recognizable. Its twenty- to forty-inch-long stems emerge from a swollen or bulblike base that forms large clumps. The "bulb" or "corm" at the base (and there may be two) is actually a swollen or thickened internode. Each "bulb" is annual in duration; it forms during the summer and dies the next year when the seed matures. Timothy, used as pasture grass, should be cut in early bloom because the food value of the grass decreases and its fiber content increases as the season advances.

The generic name comes from the old Greek *phleos*, marsh reed. The earliest mention of this grass as "timothy" was in a letter dated July 16, 1749, written by Benjamin Franklin to Jared Eliot. It is possible that the common name most popularly used for this grass comes from Timothy Hanson, who is said to have brought the grass to Maryland from northern England. By 1807 it was *the* hay grass of the American Colonies. However, it is also called Herd's grass after the man who is supposed to have found it growing along the Piscataqua River near Portsmouth, New Hampshire.

Wear gloves and pull this large grass out, or apply (checking directions on label first) Paraquat, to which the grass is moderately sensitive in lower doses and very sensitive in higher concentrations. It responds similarly to Asulam.

Timothy [*Phleum pratense*]. A. Habit of plant showing swollen bases. B. The crowded terminal, green spike (a tightly contracted panicle).

A X1

B

Setaria viridis (L.) Beauv.

COMMON NAMES Yellow foxtail grass, summer grass, golden foxtail, wild millet, pussy grass.
SOME FACTS Introduced from Europe; annual; reproduces by seeds.
BLOOMS July to September.
RANGE Common throughout North America.
HABITAT Waste places, rich soils, pastures, cultivated ground, just about any soil.

Both this plant and species of *Panicum* (see pp. 89–91) are closely related to foxtail millet (*Setaria italica*), which was cultivated in China for its grains as early as 2700 B.C., and is still cultivated there. Millets are generally a poor man's cereal. Birds and turkeys are fond of the grains, too.

Wild millet may attain a height of four feet when happy with its surroundings, and plants one to three feet high are common. The plant will even flower at a height of only a few inches if it has been repeatedly cut before flowering time. Its stems branch at their bases and bear three- to six-inch-long flat, smooth, linear-lanceolate leaves that hang with a twist. They are one-half inch wide.

The spikelets of the one- to four-inch-long spike contain one flower each and are closely packed together. Each seed is subtended by a cluster of yellowish-brown barbed bristles that are much longer than the seed. As a result, the whole inflorescence looks like a bottle brush or "foxtail." When the inflorescence has aged and dried, it may be dangerous to cattle, producing oral disturbances and infections.

To rid land of foxtail grass, use a preemergence treatment with CDAA (Randox T). Simazine has been found to be effective in preventing the germination of its seeds.

Yellow foxtail [*Setaria viridis*]. A. Growth habit of foxtail grass. B. Foxtail-shaped spike of flower.

Lolium perenne L.

COMMON NAMES Perennial ryegrass, raygrass, common darnel, English ryegrass.
SOME FACTS Introduced from Europe; perennial; reproduces by seeds.
BLOOMS June to July.
RANGE Throughout northern United States.
HABITAT Fields, meadows, roadsides, pastures, lawns.

Perennial ryegrass was grown in pure stands as a forage grass as early as 1681, and the first written mention of this grass occurs in agricultural literature as early as 1611. It is still a popular forage grass in Europe where the cool moist summers are ideal for it.

The erect culms of common darnel stand ten to thirty inches high and are smooth. Leaves are flat and smooth. The mature plant is reddish near its base.

A three- to eight-inch terminal spike of flowers is produced, on which the flattened spikelets are pressed closely against a concave space in the rachis. Each spikelet is enclosed in but a single glume and each is composed of five to ten flowers. No awns are seen on the lemmas or on the seeds.

Common darnel's close relative, poison darnel (*Lolium temulentum*), was known to one of the authors of the Bible, who was probably referring to this plant when he wrote of "tares." Theophrastus, colleague of Aristotle, knew this plant. The poison, lacking in perennial ryegrass, is produced by a fungus that attacks the poison darnel.

To destroy, hand pull, but wear gloves.

Perennial ryegrass [*Lolium perenne*]. A. Rhizome or rootstock. B. Aerial stems with clasping leaves. C. Spike with alternating spikelets.

A X1

B

Andropogon virginicus Michx.

COMMON NAMES   Little blue stem, broombeard grass, wolf grass, poverty grass.

SOME FACTS   Native to our shores; perennial; reproduces by seeds.

BLOOMS   July to September.

RANGE   Widespread in the United States, especially prevalent in prairie states.

HABITAT   Waste ground, dry fields, pastures, especially in sandy soil.

If the scientific name of this common weedy grass is translated, it is a relatively good description of some of the most striking features of the plant. In Greek *andros* means man and *pogon* beard. The inflorescences of little blue stem contain numerous filmy hairs, which give them a feathery or bearded appearance, while the overall aspect of the plant is broomlike (in an old-fashioned sense—don't compare it with the modern mass-produced broom).

Standing one to three feet tall, the flat, erect stems occur in tufts. The leaves of this long-lived native bunch grass are, at first, light green, but later characteristically became reddish. Two types of spikelets are found on little blue stem: a sessile fertile one and a stalked sterile one. The latter bears the numerous filmy hairs that give the inflorescence its bearded appearance.

Little blue stem is highly prized as a forage grass, along with its cousin *Andropogon gerardi*, which is known as big blue stem. Both form the most prevalent constituents of wild hay in the prairie states. When mature, both grasses are less palatable. Broombeard grass thrives on a wide range of soils, and since it is drought-resistant, it is useful in erosion control.

Use gloves when you pull out the bunches of little blue stem.

Little blue stem [*Andropogon virginicus*]. **A.** Flattened erect culms in tufts. **B.** Inflorescence of several spikes obscured by filmy hairs.

A

B X1

Digitaria sanguinalis Scop.

COMMON NAMES Crabgrass, fingergrass, Polish millet, crowfoot grass, pigeon grass, purple or large crabgrass.

SOME FACTS Introduced from Europe; annual; propagates by seeds and by rooting nodes.

BLOOMS July to September.

RANGE Worldwide. Widespread throughout America, very noxious in Northeast.

HABITAT Waste places, bare patches in lawns, along ditches, in cultivated fields.

If the question, "Can you name a very noxious weed?" were put to one hundred randomly chosen suburbanites, a very high percentage of them would reply, "Crabgrass!"

Fortunately, this grass is an annual, but unfortunately, the oodles of seeds it forms in only one growing season are long-lived, and wherever the stems touch the ground, the nodes root.

While it may be a noxious lawn weed, in the South it is a blessing to the farmer who wants a rapidly growing hay plant with pasturage possibilities after the hay has been collected. Strawberry growers let it fill in the spaces between the rows of berry plants where, otherwise, they would have to scatter straw.

Digitaria's culms can get as long as four feet, though usually they are shorter and decumbent or creeping at the base. Each leaf is about three to six inches long and about one-fourth inch wide.

At the top of the culm three to ten spikes of tiny flowers make their appearance and each spike is two to five inches long. The scientific name of fingergrass—*Digitaria*—comes from the handlike appearance of the clustered inflorescences at the top of the stem, and since they turn bright red to purple, blood (*sanguinalis*) is suggested.

Crabgrass germinates from seed on the lawn between midspring and late summer and will set seed even if the lawn has been mowed to as low as one-fourth of an inch! (How much lower can you get?) All crabgrass is killed by the very first frost.

Lawn grass can't take the summer heat and bright sunlight as well as crabgrass, but desirable grasses can take shading a bit better and crabgrass can't take it at all. Indeed, improper mowing is one of the major causes behind the invasion of suburban lawns by weeds.

Chemical companies that produce weed killers have been working very hard on the crabgrass problem for years now, and with relatively good results. J. M. Duich of Pennsylvania State University, an outstanding authority on turf grass management, states that preemergence chemicals should be applied "at lilac blooming time" and recommends a chemical trade-marked as Typersan because it is tolerated by turf grasses.

Crabgrass [*Digitaria sanguinalis*]. A. Branching spike with spikelets. B. New nodal roots growing out where stem touches ground.

A

Eleusine indica Gaertn.

COMMON NAMES   Goose grass, yard grass, wire grass, crowfoot grass.
SOME FACTS   Introduced to United States from warmer parts of Asia;
   annual; propagates by seeds.
BLOOMS   June to September.
RANGE   Widespread in the United States, especially in the South.
HABITAT   Lawns, roadsides, waste places, yards.

Goose grass—a serious lawn pest only when the ground is poor and hard—is most often confused with crabgrass, and it is easy to see why. The sessile spikelets, in two rows on two, are on three to eight branches that are digitately arranged at the apex of the culm. This corresponds, though only superficially, with the fruiting structure of crabgrass. However, crabgrass's inflorescence ends in a small extension of bare stem that makes the digitately displayed spikes look like a claw (see crabgrass, p. 85), while the ends of the spikes of goose grass flowers are blunt, with spikelets running right to the tips of the branches.

Below the digitate rows of spikelets there may be one or two branches with spikelets. The stem or culm of this low-spreading annual grass is thick, flat, and weak, and often prostrate. In the northern sections of the United States yard grass is usually relatively short, but in the South it may attain a height of two feet. Its leaves have loose, overlapping, flattened sheaths and flat, pale green blades.

Wire grass is cultivated in India (hence the specific epithet *indica*) for grain and forage because it is usually a heavy yielder on poor soils. Its seeds were used in poor sections of Europe as a flour source and have been recommended for a porridge during poor times. Its use as a grain source in early times is suggested by its generic name *Eleusine*, taken from the Greek city Eleusis in which a temple of Ceres, protectoress of grain, was found.

Hand pull wire grass after putting on your garden gloves. An oil can full of crude carbolic can be of some help—squirt some into the center of a tuft of the grass and it should perish.

Goosegrass [*Eleusine indica*]. A. The fingerlike spike of spikelets.

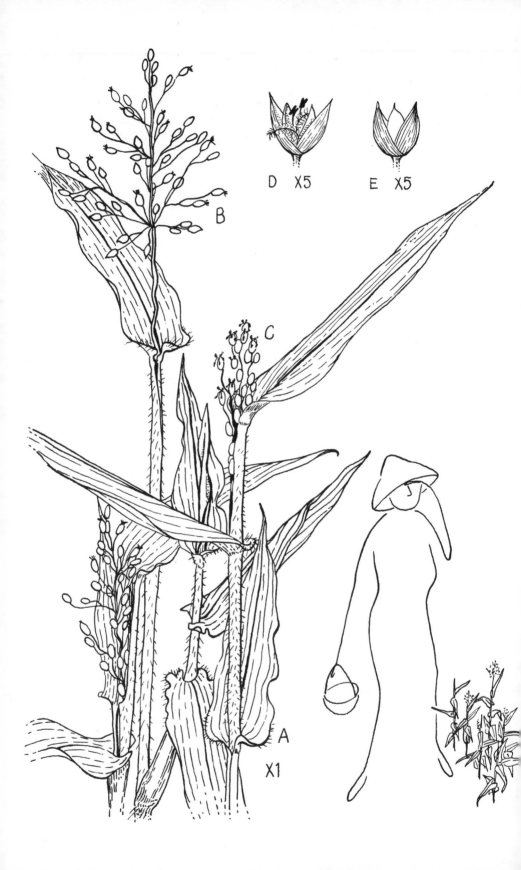

D X5   E X5

B

C

A

X1

Panicum clandestinum L.

COMMON NAMES Panic grass, deer-tongue grass.

SOME FACTS Native of the United States; perennial; reproduces by seeds and by rhizomes.

BLOOMS June to July and clandestinely.

RANGE Quebec and northern United States to Michigan, Missouri, and Oklahoma, and south to Florida and Texas.

HABITAT Moist woods, thickets, along edges of open areas.

It would be very easy on seeing *Panicum clandestinum* for the first time to fail to recognize it as a grass. With its wide, diverging leaves and sturdy culms, its obvious parallel venation of leaves that clearly clasp the stems, it is obviously a monocot, but to the untutored eye, its grass nature is not announced as loudly. Indeed, it might be thought closer to the day lily (*Commelina communis*, p. 105), a lily relative, than to *P. virgatum* (p. 91), an obvious grass.

Adding to the difficulty is the fact that its inflorescences are clandestinely produced, that is, hidden from the eye until late in the season when the mature seeds are exposed in panicles. No one could confuse the involucre, if visible, with that of a lily-type flower. Deer-tongue grass's flowers remain hidden within the clasping leaf bases of the upper leaves, emerging only after seed set has occurred. These leaves have overlapping leaf sheaths, while the culm has very much shortened internodes. One may wonder how the flowers are pollinated. What is the agent? Insect? Wind? No, the much-reduced flowers are clandestinely pollinated, that is, self-pollinated while hidden within the leaf sheaths.

This species of grass is related to *P. miliaceous*, known to us as millet, thought to be the very first of the cultivated grains.

P. clandestinum grows, usually, in large rhizome-sponsored colonies and is strongly rooted. When pulling it out, be sure to wear gloves to protect your skin from the stiff hairs in the sheaths.

Panic grass [*Panicum clandestinum*]. A. Large, flat sessile leaves; leaves wide enough to fool the seeker of grasses. B. Emergent panicle. C. Panicle often more hidden, or clandestine. D. Single flower. E. Single fruit (caryopsis).

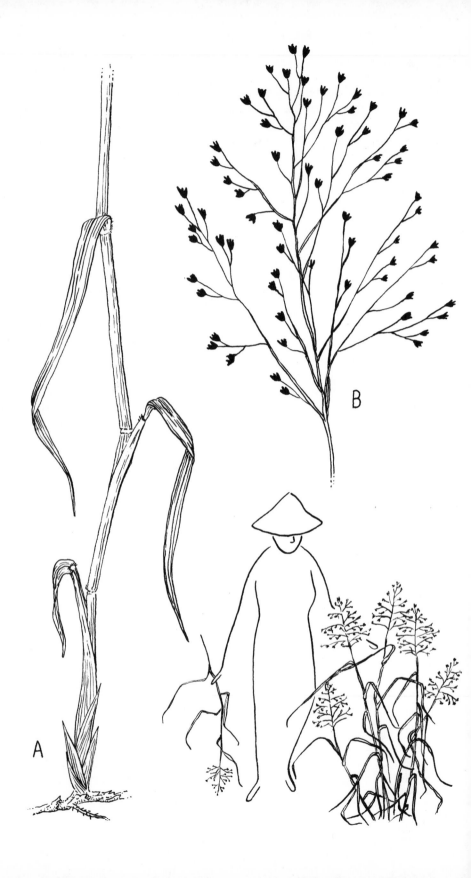

A

B

Panicum virgatum L.

COMMON NAME Switch grass.

SOME FACTS Native to the United States; perennial; reproduces by seeds and rootstocks.

BLOOMS August to September.

RANGE Widespread from Maine to Manitoba and southward to Florida and Mexico.

HABITAT Sandy soil, salt marshes along the coast, along stream banks, in low meadows.

Switch grass is a vigorous, native, perennial, sod-forming grass that occurs throughout most of the United States. It is most abundant and important as a forage and pasture grass in the central and southern parts of the Great Plains.

Its smooth stems grow to five feet under good conditions and are often glaucous. The leaf blades may be a foot long and one-half inch wide. The leaves, too, are smooth and often bluish-green in color. While their surfaces are smooth and flat, their margins are slightly rough to the touch.

Erect six- to twenty-inch-long panicles that are spreading and pyramidal are found on this grass. The spikelets are one-seeded and ovate.

Though this weedy grass is less often found in gardens than along roadsides, it should be removed when it makes its appearance among cultivated plants. CDAA (including Randox T) and simazine may be used for preemergence treatment; otherwise pull the grass out (wear gloves).

Switch grass [*Panicum virgatum*]. **A.** Tall culm showing leaves. **B.** Open panicle.

A

B

Tridens flavus (L.) Hitchc.

COMMON NAMES Tall redtop, purpletop.

SOME FACTS Native to North America; perennial; propagates by seeds and by rootstocks.

BLOOMS July to September.

RANGE New Hampshire to Nebraska, south to Florida and Texas.

HABITAT Old fields, open woods, waste grounds, and gardens.

Tall redtop is certainly striking the first time you happen on it in a field. Students are particularly impressed if they happen on it early in a semester of field work because they believe all grasses are short and flat. Students also like it later in the semester, when the time comes to prove they can identify some plants, because *Tridens flavus* is easy to recognize in flower.

The most striking feature of redtop is, of course, its quite large, open panicle with its drooping branches (most characteristically, the lower ones). The panicle turns a bright red-purple when mature and later exudes a sticky substance that collects dust and grime from the air. If you squeeze the spikelets, you can feel this stickiness. On an occasional rare plant the panicle may be yellow (*flavus* is Latin for yellow—apparently when first seen, the rare one was observed).

This grass may grow from two to five feet high and is composed of several slender culms and their associated leaves—often a foot long—from a several-tufted base.

Pull (with gloves on) wherever you object to its presence.

Tall redtop [*Tridens flavus*]. A. Tall culm ending in (B) the large, open panicle, purplish-red in color.

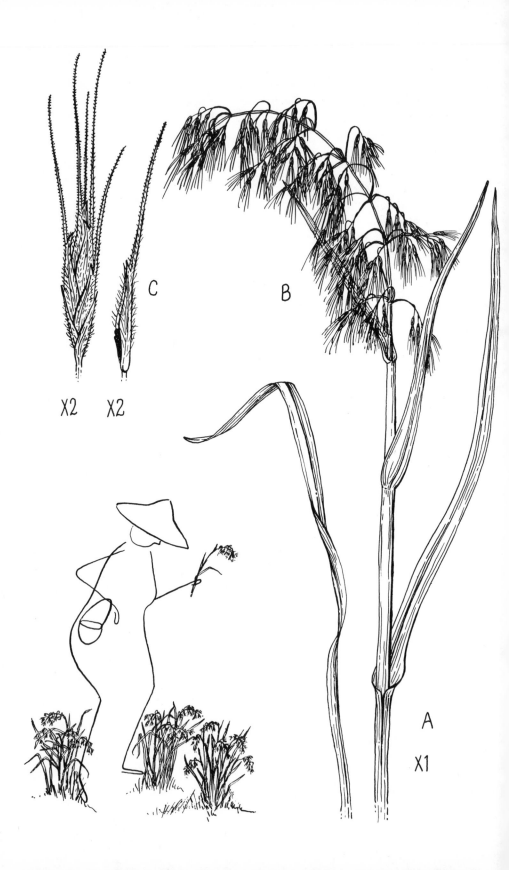

C   B

X2   X2

A

X1

Bromus tectorum L.

COMMON NAMES Early chess, downy bromegrass, slender chess, downy chess, cheatgrass.

SOME FACTS Annual or winter annual; native of Europe; reproduces by seeds.

BLOOMS May to July.

RANGE Widespread in United States except in the Southeast.

HABITAT Roadsides, waste places, fields, especially on dry, sandy, or gravelly soils.

Downy bromegrass is a beautiful sight at maturity. The drooping, dense panicles of long slender spikelets look like purple plumes as they nod in the wind at the tops of the one- to two-foot-long stems. Its leaf blades and sheaths are very pubescent.

Remove manually before seed dispersal.

Downy bromegrass [*Bromus tectorum*]. A. Culm, showing clasping leaf. B. The characteristic drooping panicle. C. Individual florets showing awns. The power of magnification is denoted by X.

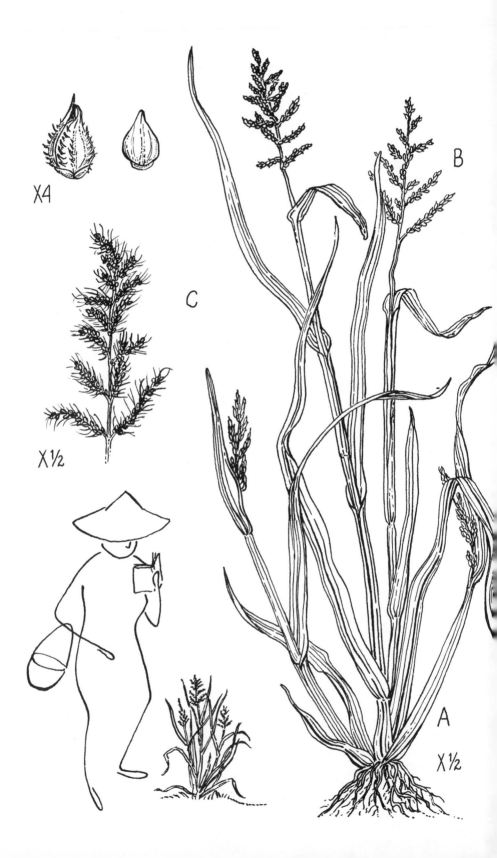

X4

B

C

X½

A

X½

Echinochloa crusgalli (L.) Beauv.

COMMON NAMES Barnyard grass, cockspur grass, water grass, cocksfoot panicum.
SOME FACTS Coarse annual; from Eurasia.
BLOOMS July to September.
RANGE Cosmopolitan.
HABITAT Moist ditches, manured soils, cultivated fields, gardens, rice fields.

Now that only 7 percent (or fewer) of our people farm to feed the other 200,000,000 of us, there are few people who have seen this weedy grass where it used to be omnipresent—in the barnyard. However, this grassy weed is not limited to the barnyard; it also invades the garden. It has been reported that its seeds survive for nine years in the soil, a troublesome thought for the gardener.

Yet young barnyard grass is useful and can be fed to cattle, which seem to find it quite palatable. The Indians of Arizona and southern California have used the seed for cattle feed for a very long time. It is eaten as a famine food by the Chinese.

Echinochloa is a tall plant with stout three-foot-tall culms that may be somewhat decumbent. It has long leaves and each has a pale midrib. However, it is its inflorescence that makes it so easy to identify. It has large four-inch-long panicles that are easily spotted because the branches of this spreading panicle are fringelike and bear long bristles. Each bristle is a barbed awn associated with a flower. Its generic name comes from this bristly spreading panicle at the top: *Echino-chloa* means green hedgehog and *crusgalli* means cock's foot. Sometimes the inflorescence is a deep purple rather than a pale yellow.

Barnyard grass is considered particularly bothersome in rice fields in the United States, and causes much annoyance in carrot fields and fields with other root crops. It is listed in Dr. Alden Craft's *Modern Weed Control* as one of the world's ten worst weeds.

It is a difficult weed to get rid of and postemergence treatments with herbicides are not very successful. However, preemergence treatments with CDAA and simazine seem helpful. Exercise special care in using herbicides of any kind.

Barnyard grass [*Echinochloa crusgalli*]. **A.** Whole plant. **B.** Flowering stalks. **C.** Open panicle.

A

Dactylis glomerata L.

COMMON NAMES Orchard grass, cocksfoot.
SOME FACTS Introduced from Europe; perennial; reproduces by seeds.
BLOOMS July to August.
RANGE Common throughout North America except in desert or arctic regions.
HABITAT Open fields, pastures, lawns, gardens, roadsides.

Orchard grass is one of the first grasses to grow in the spring, and among the last to give in to the frosts of fall. This larger member of our grass family was introduced to America in 1760 as a desirable pasture grass. It had been cultivated for centuries in Europe as a meadow and pasture grass. It is not seriously bothered by diseases.

Cocksfoot is a tall, erect grass that grows as high as three or four feet. It grows in great clumps rather than in dense sods. This is because of its failure—Zeus be thanked!—to produce rhizomes. The large tussocks formed from the densely clustered culms are topped in height only by the large dactyloid inflorescence. The densely crowded spikelets are in irregular clusters at the end of the branches of the large, tall panicle.

This bunch-type grass with folded leaf blades is but moderately sensitive to sodium chlorate, or to boric oxide or simazine. Since its rootstocks do not spread, pulling out the young plants is wise, but wear your garden gloves.

Orchard grass [*Dactylis glomerata*]. A. Densely crowded spikelets in irregular clusters at end of the branch of the panicle.

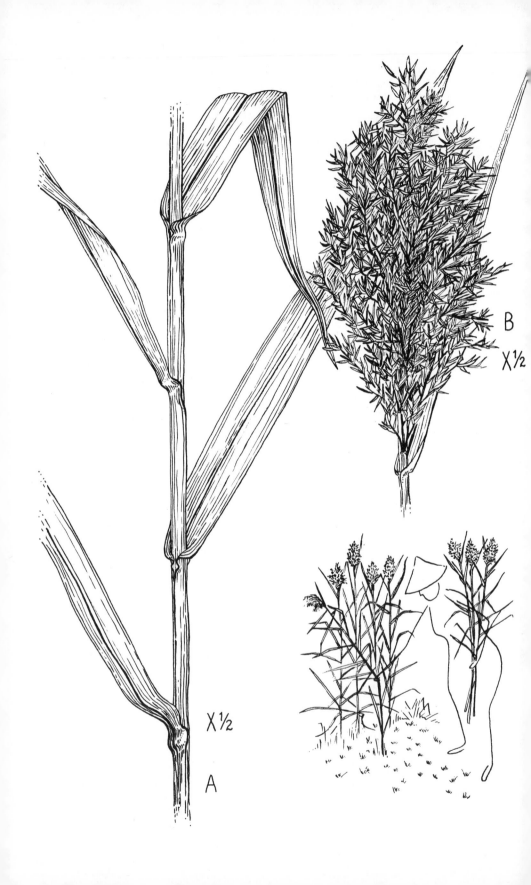

B

X½

X½

A

Phragmites communis Trin.

COMMON NAMES Reed grass, cane grass.

SOME FACTS Introduced from Europe; perennial; rarely propagates by seeds.

BLOOMS June to July.

RANGE Nova Scotia to British Columbia, then south through the United States; also in South America and Australia.

HABITAT Drainage channels, shallow water, marshy wet shores, along ditches (next to roadways).

It is hard to imagine that anyone with an interest in plant life has not seen *Phragmites*, one of the tallest grasses in our country. Its culms attain a height of six to nine feet under marshy conditions. Topped by a puff of delicate flowers, these culms are often gathered and placed in vases for their aesthetic qualities. The nodding heads of flowers are a conspicuous dull purple.

The generic name of this plant comes from the Greek *phragma*, which means a fence or screen, and the impression of a reed screen is definitely produced by the tall, thin stems, which resemble slim bamboo stalks. The erect stems arise from underground creeping rhizomes.

Reed grass's older stems have been used to thatch roofs, and the young shoots may be either boiled and cooked as asparagus or pickled. The rhizome from which the aerial shoots arise can be boiled and eaten just like a potato.

To destroy any plants of reed that have, perhaps, crept in from a drainage ditch on or near your property, use Dalapon (15 to 20 pounds per acre). Be sure to apply it when the shoots are growing vigorously and are about two to three feet tall.

Reed grass [*Phragmites communis*]. **A.** Stout, tall culm of reed grass showing clasping, parallel-veined leaves. **B.** Plumelike mass of spikelets.

C

D

B

X ½

A

X ½

Other Monocots

Hemerocallis fulva L.

COMMON NAMES   Day lily, tawny-orange lily.
SOME FACTS   Introduced from Europe; perennial; seeds only very rarely, usually propagates from tuberous roots.
BLOOMS   June to July.
RANGE   Escapes from gardens locally; common in northeastern United States and in other parts of world.
HABITAT   Cultivated soils, roadsides, backs, waste places, usually in rich, damp, gravelly soils.

Day lily escapes easily from cultivation and soon covers an area on which other plants may be wanted, not by producing seeds, but by increasing its tuberous roots. Indeed, the three-chambered capsule, when it does develop, usually does not contain viable seeds.

From the tuberous roots arise long, linear, two-ranked smooth leaves, and from a naked, three- to four-foot simple stem several largish orange-colored typically lily-shaped flowers take form and bloom one at a time. Each flower, composed of six bright orange perianth parts, lasts but a day (Greek *hemera*, one day, and *kallos*, beauty).

One of the common names for this plant is tawny-orange lily, and the botanical name also reflects its color: *fulvus* in Latin means orange. The day lily's orange unopened flower buds may be prepared in a number of ways and are delicious. The young tuberous roots, when boiled, are very tasty.

While day lily is basically a beautiful plant, it tends to run wild and should be checked. Dig out the roots. Spraying the young leaves with sodium chlorate may also do the trick.

Day lily [*Hemerocallis fulva*]. **A.** Rhizome, which establishes new plants. **B.** Top of tall (three to five feet) stalk with flowers. **C.** Unopened flower bud. **D.** Large flower with three sepals looking like the three petals, six stamens, and three fused carpels (fused into one pistil).

Commelina communis L.

COMMON NAME Dayflower.

SOME FACTS Native of Asia; annual; reproduces by seeds and by creeping stems.

BLOOMS July to September.

RANGE Massachusetts to Florida, west to Kansas and Texas.

HABITAT Gardens, neglected fields, waste-rich soil.

Commelina, the dayflower, is actually quite attractive, but can take over a shady section of the garden so rapidly that other equally or more attractive plants may not have a chance to survive. The speed of travel of this pretty weed comes from its creeping stem, which roots rapidly at its swollen nodes.

One of our very few weeds from Asia, its leaves—two to four inches long—are alternate, simple, and entire, with the parallel veins expected in a monocot. *Commelina* is a member of the spiderwort family, to which belongs the popular garden plant *Tradescantia*, whose common name is widow's tears.

The genus *Commelina* was named by Linnaeus, who took his cue from the presence of two lateral bright blue petals and a very much smaller and paler (or white) third petal beneath the two lateral ones. Let the great taxonomist of the eighteenth century tell you in his own words: "*Commelina* has flowers with three petals, two of which are showy, while the third is not conspicuous; from the two botanists called Commelin, for the third died before accomplishing anything in Botany." The Commeljn brothers were Dutch botanists.

There are also six stamens, only three of which are fertile (three are fused pistils).

Pull stalks up. If new stems and leaves appear from the rootstock, pull again. Repeat until rootstocks have been killed.

Dayflower [*Commelina communis*]. A. Habit sketch of plant. B. Relatively wide and short leaves (for a monocot) with parallel venation and clasping bases. C. Single flower emerging from protective basal leaves. D. Flower, showing two larger (blue) petals and single smaller petals. E. Later, capsules with their four or five brown seeds. The power of magnification is denoted by X.

HERBACEOUS
PLANTS

Dicotyledonous Species

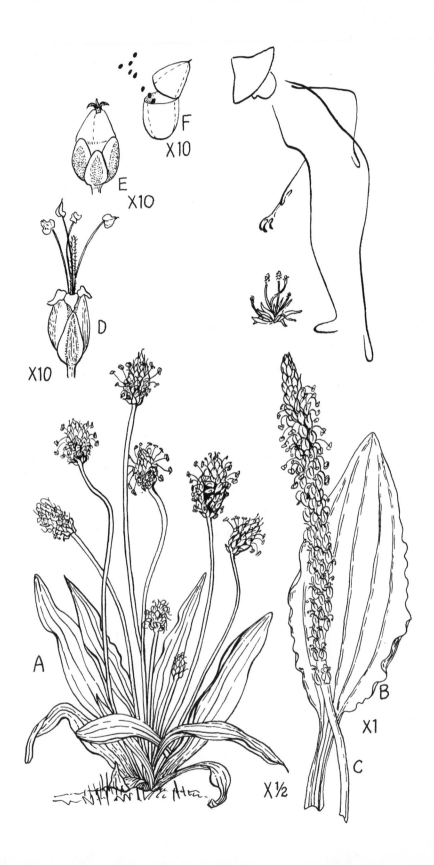

A

B

X1

C

D

X10

E

X10

F

X10

X½

Green-Flowered Species

Plantago lanceolata L.

COMMON NAMES English plantain, ribgrass, buckhorn, ribwort, narrow-leaved plantain, ripple grass, blackjacks.
SOME FACTS Introduced from Europe; perennial; propagates by seeds.
BLOOMS May to October.
RANGE Throughout the United States and Canada.
HABITAT Lawns, roadsides, waste places.

Ribgrass, noxious as a weed, has been very useful in reading the early history of civilizations in Europe. The plant grows, as most weeds do, in open places and has been used by ethnobotanists in a very interesting way.

When neolithic people cleared the forests of Europe, often by burning them, they planted their simple crops for a few seasons and then, the land exhausted or for other reasons, moved on to new ground. As soon as they cleared the land, ribgrass sprang up all around the simple farms and, no doubt, on the farmland itself, though I suppose we can assume that those early farmers tried to root out these weeds. The pollen shed by ribgrass fell to earth and was covered, year by year, by layers of dust and soil. When the early farmers moved on to a more fertile area, the now neglected farmland first became a meadow with tall weeds. Into the meadow moved, eventually, taller weeds and, later still, shrubs. Finally the land reverted to forest. These changes are part of the standard cycle of plant succession —from forest to farmland to meadow and later back to forest again, after human beings have departed. Our cities would go the same way were we to leave them. This cycle can be read in the layers of soil with their various pollens, and the pollen of ribgrass is especially helpful in reading these stories because ribgrass is such an excellent barometer of the changes in the openness of the land.

Though the dark green, thin, lanceolate leaves (hence its specific epithet) are deeply ribbed by what seem to be parallel veins, do not make the erroneous assumption that ribgrass is a monocot. It is a member of the dicots. This crown of lanceolate, ribbed leaves sits on a very short stem (not really observable from above) from which emerge many fibrous roots.

The flowering scapes of English plantain are different from those of the broad-leaved plantain (*P. major*) in that the flowering portion of the scape is contracted into a short, dense spike, which sits on a long thin, angled stalk. The spikes flower from the bottom, as do those of the broad-

---

Ribgrass [*Plantago lanceolata*] and common plantain [*P. major*]. A. Whole plant of ribgrass showing flowering stalks and ribbed, lanceolate leaves. B. Single leaf of common plantain showing ribbing and its greater width. C. Longer flowering stalk of *P. major*. D. Single staminate flower with long stamens. E. Single flower. F. Fruit (capsule), which opens around equator (circumscissile). Note tiny seeds.

leaved relative. The stamens, sticking out from the dense greenish-brown spikes, look like pins in a pincushion.

Pulling the English plantain before it comes into flower may not entirely banish this lawn pest because the many fleshy roots can give rise to new stems. If the top is removed and sodium chlorate or carbolic acid placed in a hole pierced through the exposed tissue, the plant should then be doomed.

Plantago major L.

COMMON NAMES   Common plantain, greater plantain, dooryard plantain, white man's foot, broad-leaved plantain.
SOME FACTS   Introduced from Europe; perennial; reproduces by seeds.
BLOOMS   May to September.
RANGE   Throughout United States and Canada.

The common name white man's foot was given to *Plantago major* by the American Indians. Peter Kalm attested to this in his journal in a notation made on September 26, 1748: "The broad plantain, or *Plantago major*, grows on the highroads, foot paths, meadows and in gardens in great quantities. Mr. Bartram [John Bartram, an early American botanist at Philadelphia] had found this plant in many places on his travels, but he did not know whether it was an original American plant or whether the Europeans had brought it over because the Indians (who always had an extensive knowledge of the plants of this country) pretended that this plant never grew here before the arrival of the White Men. They therefore gave it a name which signified [Englishman's] foot, for they say that 'wherever a European had walked, this plant grew in his footsteps.' "

P. major had followed the Englishmen from Europe, but *P. rugellii* was here when he arrived. The plants look very much alike but differ in two quickly distinguishable ways: The color of the petioles (those of the common plantain, *P. major*, are green, while those of *P. rugellii* develop a deep red-purple pigment) and the position of the line of dehiscence (opening) of the fruits.

No unattended lawn is safe from the depredations of either species. Their rosettes of ellipsoidal, often spoon-shaped, deeply veined leaves quickly shade out more desirable lawn grasses. Unattended lawns may soon be filled with plantains. Numerous seeds are produced within each tiny capsule and there are many capsules formed along the three- to twelve-inch-long, rattail-shaped spikes of flowers that arise from the center of each rosette. A single, tiny flower is composed of four sepals, four petals, and four stamens with long filaments that stick out of the flower. The capsules, formed after fertilization of the flower, open transversely by a lid. (In *P. rugellii* the capsules open by means of a break that appears horizontally all around the fruit, while those of *P. major* open at approximately the equator of the fruit.)

No weedy species of plantain are put to use by man, but the seeds of the very closely related *P. psyillum*, *P. ovata*, and *P. indica* are used to overcome the sluggishness of man's large colon. John Gerard, the ancient herbalist, has something to say about the usefulness of *Plantago*: "The juice dropped in the eies colles the heate and inflammation thereof. I find in antient writers many good morrowes, which I thinke not meet to bring into your memorie againe; as, That three roots will cure one griefe, foure another disease, six hanged about the necke are good for another malady, etc., all of which are but ridiculous toyes." I think he has something there.

Fortunately, the broad-leaved plantain does not root deeply beneath its almost stemless crown of leaves, and the plants can be pulled up with relative ease. Pull up before the flowering scape matures. Any strong acid such as carbolic will kill these perennials. A single application of MCPA salt (twelve ounces per acre) or 2,4-D amine (twenty-four ounces per acre) should restore the evenness of your lawn.

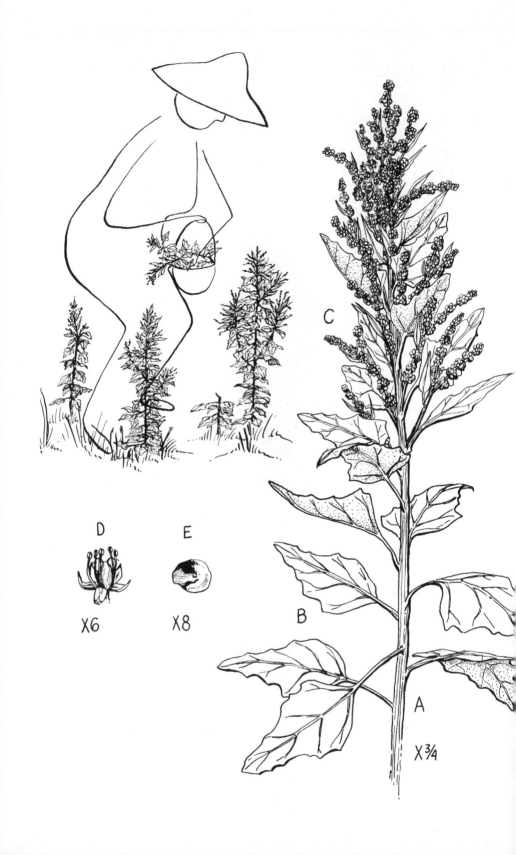

D
X6

E
X8

B

C

A
X¾

Chenopodium album L.

COMMON NAMES White goosefoot, lamb's-quarters, pigweed, fat hen, meal-weed, frost-blite, bacon weed.

SOME FACTS Introduced from Eurasia; annual; reproduces by seeds.

BLOOMS June to September.

RANGE Common throughout the United States.

HABITAT Waste ground, cultivated fields, gardens.

It is possible to bake a palatable bread using the seeds of mealweed. Occasionally, during politically stressful moments, Napoleon I himself had to live on chenopodium bread. American Indians used it for bread and gruel. The seeds may not only be mixed with wheat and baked into bread, but may also be boiled and eaten or toasted as tortillas. Goosefoot's leaves and stems make excellent salad greens if young enough, and a very popular potherb cooked in the manner of spinach or the tops of beets; both spinach and the beet are closely related to goosefoot.

Proof of its ancient popularity has been found in long dead but completely preserved human stomachs. During the third to fifth centuries A.D., it was not uncommon among some northern European peoples to bury their dead in the peat bogs. There they remained until scientists dug them out. Since bogs are highly acid, these long-dead Continentals never corrupted but were, instead, tanned as leather is tanned. Analysis of the contents of their stomach reveals that goosefoot seeds were a popular food with them.

Lamb's-quarters grows very rapidly under good conditions and may attain a height of six feet. Its sturdy stem is ridged and grooved and usually much branched. Stems of older specimens may be striped with reddish purple.

Near the base of the stem are seen alternate petiolate leaves that are rhombic-ovate or goosefoot-shaped. Higher up on the stem the leaves are narrower and lanceolate, and at the top of the plant they become linear and sessile. The leaves are dark green when viewed from above and whitish or light gray-green and mealy when viewed from below.

Small green flowers are crowded on irregularly spiked clusters in panicles found within the axils of leaves and at the tip of the plant. The five lobes of the mature calyx enfold the lens-shaped dark-colored seeds, each of which has a marginal notch. These seeds have been shown to be very long-lived.

Fat hen is sensitive to MCPA salt, 2,4-D amine, MCPB salt, and many other chemicals. Use your gardening gloves if you decide to remove pigweed by pulling.

Lamb's-quarters [*Chenopodium album*]. A. Portion of the sturdy ridged stem. B. The alternate rhombic-ovate or goosefoot-shaped leaves. C. The irregular spiked clusters (contracted panicles). D. Much-enlarged flower. E. Single lens-shaped seed.

A   X¾   B   X3   C

Xanthium pennsylvanicum L.

COMMON NAMES Cocklebur, clotbur, sheep bur, button-bur, ditch-bur.

SOME FACTS Native to Eurasia and Central America; annual; reproduces by seeds.

BLOOMS August and September.

RANGE Widespread from southern Canada throughout the United States to Mexico; especially troublesome in the Mississippi River Valley.

HABITAT Roadsides, fields, waste places; especially happy on riverbottom land.

Cocklebur always reminds me of the early grades of grammar school. We were terribly socially unaware then and never once picketed our kindergarten teacher, but instead threw cockleburs into the long tresses (the longer the better) of sweet little girls. Since these inch-long burs are covered with hooked spines, they clung with just the right tenacity for our evil purposes.

Clotbur may grow four feet high, though most specimens you will find will be between one and two and a half feet tall. Its rough, thick, angled stem produces, on long rigid petioles, alternate oval- to heart-shaped largish-toothed leaves. These are bristly to the touch on both surfaces.

Small green male flowers are found in short terminal spikes somewhat reminiscent of those of a related composite, ragweed (see p. 119). Lower down are found the axillary fertile heads of two pistil-bearing flowers encased in an involucre. The latter matures as a hard cocklebur containing two seeds (achenes). It is said that only one seed of the two in the bur germinates the first season.

Western Indians ground the seeds with cornmeal and squash seeds, yet these seeds are thought to be poisonous. The seedling, with its two lanceolate cotyledons, is definitely poisonous to livestock, especially pigs. The poisonous principle has been identified as a hydroquinone.

Cocklebur is weakly rooted and can easily be pulled up. Do this *before* the burs are mature. It is sensitive to MCPA salt and to 2,4-D amine.

Cocklebur [*Xanthium pennsylvanicum*]. **A.** Thick, angled stem. **B.** Oval to heart-shaped toothed leaves with ridged petioles. **C.** Ripe bur formed from two pistillate flowers encased within an involucre (modified leaves), which develops armament. The power of magnification is denoted by X.

F   X3   E   X3

D

C

B

A   X½

Ambrosia trifida L.

COMMON NAMES Wild hemp, giant ragweed, great ragweed, horseweed, bitterweed, kinghead, crownhead, horsecane.

SOME FACTS Native to the United States; annual; reproduces by seeds.

BLOOMS July to September.

RANGE Nova Scotia to Florida and west to northwestern United States, including Nebraska, Colorado, and Arkansas.

HABITAT Moist, rich soil, fields, waste places.

If three weedy plants had to be recommended for complete annihilation, I suspect that *Ambrosia trifida*, its sister, *A. artemisiifolia* (p. 119), and poison ivy (*Rhus radicans*, p. 61) would be chosen, with crabgrass and probably goldenrod as close runners-up (though goldenrod is perfectly harmless to man despite the rumors).

The goldenrods share the blame with the ragweeds for torturing hayfever sufferers but are fundamentally guiltless. To cause hayfever, a plant must produce pollen that is windblown, as ragweed does. This is not true of goldenrod. (Recently an advertisement in the New York City subways showed a lovely cluster of goldenrod flowers next to a jar of X. The goldenrod flowers do *not* produce hayfever; let us hope that the jar of X can cure the hayfever that ragweed *does* produce.)

Giant ragweed is a member of the great tribe of composites and grows to a very great height. When all its requirements are met, it may attain a height of twelve to fifteen feet. It is a large, coarse plant that will crowd out cultivated plants in the garden or shade out other weeds in an open lot.

Though many of its leaves are trifid, that is, have three major lobes, many others on the plant have five lobes and the younger ones near the top of the plant are unlobed. Individual leaves on taller plants may attain a length of one foot. All are coarsely toothed and have stout petioles.

The staminate flowers shed pollen abundantly. The fruit that forms following pollination is about one-half inch or more long, brown in color, and has five or six ribs. A conical beak is found at the apex surrounded by five or six shorter spinelike structures that look like the points of a crown; hence the common name crownhead or kingshead.

An application of 2,4-D should help rid you of this plant pest. Hand pulling of larger plants may not get all the deep roots, but will temporarily end your problem; watchfulness and attention should help you catch any new shoots.

Giant ragweed [*Ambrosia trifida*]. A. Large, rough-hairy, stout stem. B. Large rough-hairy, trifid (deeply three-lobed) leaf. C. Female flower in axil of leaves. D. Spikes of male flowers. E. Single enlarged male flower. F. Mature fruit (achene) within bracts. The power of magnification is denoted by X.

C

D

E

X1

X2

X2

♀

Ambrosia artemisiifolia L.

COMMON NAMES Hayfever weed, common ragweed, hogweed, bitterweed, wild tansy, roman wormwood, carrot weed, stammerwort.

SOME FACTS Native to the United States; annual; reproduces by seeds.

BLOOMS July to September.

RANGE United States and Canada, from Nova Scotia to British Columbia south to Florida and Texas.

HABITAT Dry soil, cultivated ground, waste places.

It is ironic, indeed, that this most noxious rank weed, this cause of so much late-summer human suffering, should be called *Ambrosia*. Apparently, what is food for the gods is not good for man. Hogs and sheep, however, love the common ragweed, which accounts for its other common name, hogweed. Hayfever weed is the more appropriate name for this plant, which is occasionally mistaken for A. *vulgaris* because of the overall similarity of their leaves.

The twice pinnatifid, parted two- to five-inch leaves look almost feathery on this one- to five-foot-high cousin of the taller A. *trifida*, the giant ragweed (p. 117). Its leaves are deep green above and much lighter green below, and are alternately distributed on a stem that may get as high as five feet or remain as short as one foot. Some of the leaves near the base of the stem are oppositely arranged.

The unattractive flowers are either male (staminate) or female (pistillate). Both are found on the same plant. It is the male flower that catches your attention on both species of *Ambrosia*; these flowers are displayed at the top on prominent spikelike racemes, while the pistillate flowers lower down are concealed behind clustered bracts. Hundreds of racemes of male flowers ensure a prodigious scattering of pollen and in the process trouble for hayfever sufferers.

The growth of this plant can be controlled with 2,4-D. Wear gloves to pull individual plants out. Ragweed has deep, branching roots, so mere pulling may not rid you of this noxious weed. Repeated chemical and mechanical control will conquer it in the end.

Common ragweed [*Ambrosia artemisiifolia*]. A. Often much-branched stem. B. The rank-smelling, twice pinnatifid (feathery) leaf (the specific name *artemisiifolia* means leaves like *Artemisia*, p. 121). C. Female flower in axil of leaf. D. Maturing seed. E. Portion of male-flowered spike. The power of magnification is denoted by X.

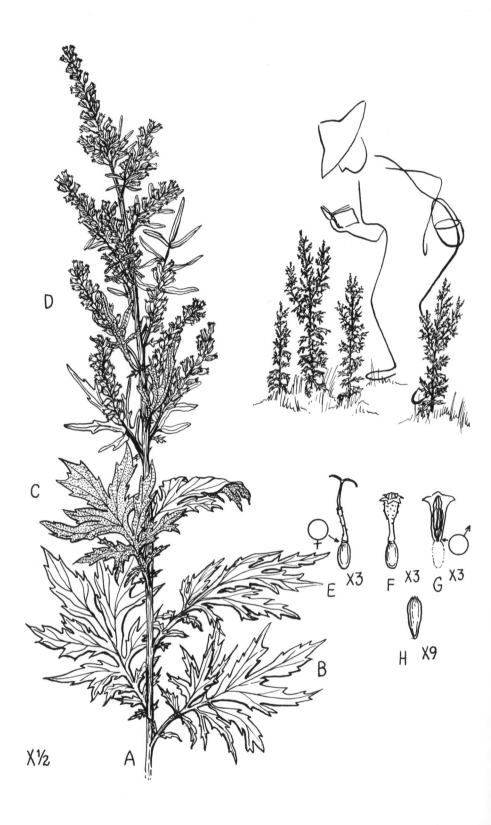

X½

A

B

C

D

E X3 F X3 G X3

H X9

Artemisia vulgaris L.

COMMON NAMES   Mugwort, wormwood, felon herb, sailor's tobacco.

SOME FACTS   Perennial; reproduces by seeds and rootstocks; may be native of north-central Asia, or of North America.

BLOOMS   July to October.

RANGE   Across northern and western United States and common in eastern United States and Canada, as well as on Pacific Coast.

HABITAT   Waste places, fields, pastures, vacant lots, often on limy soils.

Artemis was the ancient Greek goddess of the moon, wild animals, and the hunt, and was Apollo's twin sister. She was also the goddess of chastity. Unfortunately, there is nothing chaste about mugwort. It reproduces vigorously and all its small gray-green flower clusters, composed of small heads, are quite fertile and actively so.

The three-foot stems are erect and branching. Sometimes the shoots are too heavy for their own support and bend over. Though usually green, it is not unusual to see representatives that are reddish later in the season. The ridged stems come up from a perennial rhizome.

Along the stem are the alternate, deeply pinnatifid, and cleft green leaves. They are shiny green above and woolly-white below. When crushed, they give off a bitter pungent smell, not too different from that of cultivated chrysanthemum leaves. Indeed, sometimes mugwort is mistaken for the cultivated chrysanthemums and permitted to grow.

The flowers are found in small numerous heads that occur in terminal or axillary spikes, though at first glance they look more like racemes. The outer flowers of each small head are pistillate, and all are quite fertile. Following fertilization, a small oblong achene makes its appearance.

A principle derived from one of the species closely related to mugwort is used in making absinthe, a drink dangerous to consume. (Oscar Wilde, the dramatist, was very fond of absinthe.) Mugwort's leaves and flowers were used to flavor beer before hops took their place.

A famous relative in our own western lands is the sagebrush, *Artemisia tridentata*.

Said Gerard in 1597 of mugwort: "It cureth the shakings of the joynts, inclining to the palsie, and helpeth the contraction or drawing together of the nerves and sinewes." I'll raise a glass of absinthe to that!

Dig out the plant, being sure to go after the rhizome.

Wormwood [*Artemisia vulgaris*]. A. Flowering stalk showing the alternate leaves. B. The deeply pinnatifid and cleft leaf. C. Undersurface, which is woolly-white. D. Racemelike flower clusters (of heads). E. Single pistillate flower. F. More central staminate flower with petals intact. G. Staminate flower with petals removed. H. Single, much enlarged achene.

A

C

X5

X½

B

Yellow-Green—Flowered Species

Rumex crispus L.

COMMON NAMES Curly dock, yellow dock, sourdock, narrow-leaved dock.
SOME FACTS Introduced from Europe; perennial; reproduces by seeds.
BLOOMS June to September.
RANGE Throughout the United States and adjacent Canada.
HABITAT Roadsides, waste grounds, fields.

> Nettle out, Dock in
> Dock remove the Nettle sting!

Merely repeat this incantation, reported in Chaucer, while rubbing dock leaves over the region of your skin recently insulted by the needles of nettle, and the ache should leave. Curly dock's wavy-edged, lance-shaped, longish leaves, borne near the top of the long taproot, are on longish petioles. Their bases are roundish or heart-shaped.

The one- to four-foot stem is smooth, ridged, and has somewhat swollen nodes. The inflorescences appear at the top of the stem, and where they appear, branching is seen. The region of the inflorescences is usually branched several times but the branches are compacted, crowding together the drooping spikes of small six-sepaled greenish-yellow flowers.

Later in the season the stalk dies and turns a bright red-brown, each three-angled fruit maturing with the three inner sepals covering it. Indeed, in winter *Rumex crispus* can be identified from a distance by its dead red-brown stalks topped by their branched inflorescences. These are frequently used in late fall–early winter dried flower arrangements.

Both *Rumex* and *Polygonum* are members of the buckwheat family and the flour made from yellow dock's seeds is comparable to flour made from those of *Fagopyrum*, the buckwheat. The young leaves forming at the top of the taproot are used in salads, and the older ones, after several boilings, are used as potherbs. Peter Kalm recorded in 1748, "*Rumex crispus* L. is a kind of sorrel which grows at the edge cultivated fields and elsewhere in rather low lands. . . . Farmers choose a variety which has green leaves, for the leaves of some are very bitter. . . . But they generally boil the leaves in the water in which they cooked meat. Then they eat it alone with the meat. . . . I must confess that this dish tastes very good."

The narrow-leaved dock is included in an Act of Parliament, mentioned under Canada thistle (p. 225). If you find this plant moving onto your land, MCPA salt (sixteen ounces per acre) and 2,4-D amine (twenty-four ounces per acre) should rid you of it. If the thick, long taproot is well established, it may be wise to cut off as much as you can from above and put table salt on the remaining portion.

Curly dock [*Rumex crispus*]. A. Single leaf from rosette showing curly margin (hence curly dock). B. Top of long flowering stalk, with fruits shown here. C. Single fruit (achene) in chamber formed by sepals coming together. The power of magnification is denoted by X.

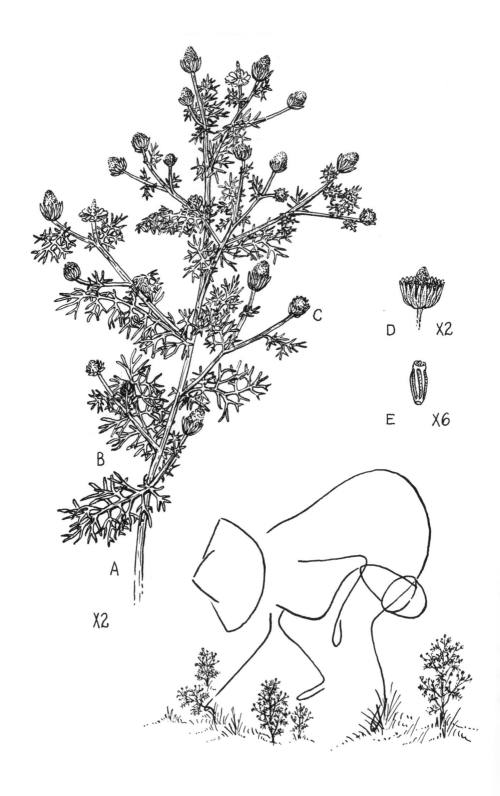

A

B

C

D X2

E X6

X2

Matricaria matricarioides (Less.) Porter

COMMON NAMES Pineapple weed, mayweed, rayless chamomile.
SOME FACTS Annual; native to the Pacific slopes; reproduces by seeds.
BLOOMS May to September.
RANGE From Alaska to Baja California and east to Montana and Arizona.
  Now in the Atlantic states, especially near cities and towns.
HABITAT Roadsides, waste places, fields.

Crush a bit of leaf material of pineapple weed and you will quickly learn why this epithet has been applied to this plant. Its pineapple-scented, finely divided leaves are one to two inches long and may be found to be one to three times pinnatifid. Even without smelling, it may be immediately distinguished from *Anthemis cotula*, the stinking daisy, which it resembles, because pineapple weed lacks any sign of ray flowers, which the odoriferous daisy does have. Each floral head is a bluntly ovoid disc, greenish-yellow in color. Many are produced on each plant.

Its name, *Matricaria*, derives from its having once been found useful in treating infections of the uterus (*mater*, mother, and *caries*, decay).

Pineapple weed resists many standard weed killers such as 2,4-D and MCPA, but is moderately sensitive to sulphuric acid applied to its foliage. It is somewhat sensitive to a combination of MCPA and 2,3,6-TBA salts.

Pineapple weed [*Matricaria matricarioides*]. **A.** A branch of pineapple weed. **B.** The pinnately dissected leaves. **C.** Small greenish-yellow head of disc flowers only. **D.** Much enlarged oblong achene with tiny crown (pappus). **E.** Single fruit. The power of magnification is denoted by X.

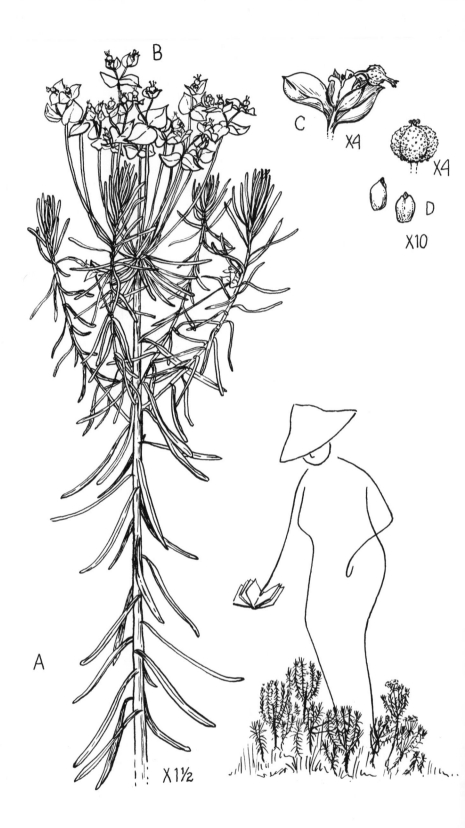

A

B

C X4

X4

D

X10

X 1½

Euphorbia cyparissias L.

COMMON NAMES Cypress spurge, salver's grass, quacksalver's grass, grave-yard weed.

SOME FACTS Introduced as an ornamental from Europe; perennial; reproduces by seeds and creeping rootstocks.

BLOOMS May to June.

RANGE Widespread and locally very common in the northeastern states and north central states; occasionally on Pacific Coast.

HABITAT Gardens, fields, waste places, roadsides, in older cemeteries.

Surely everyone knows the ever-seen-at-Christmas poinsettia (not poinse*tt*a), or *Euphorbia pulcherima*, with its huge, striking red "flowers." Actually, the real flowers are quite small, yellow-green, and unstriking; it is the red leaves (bracts) just below the cluster of tiny flowers that produce the appearance of a huge colorful bloom.

The flowers of *Euphorbia* are very reduced, having no calyx or corolla. A cup, called an involucre, which is top-shaped, surrounds several male flowers—each of which has but one stamen—and a single central female flower composed of a three-lobed pistil. Each cluster has greenish-yellow heart-shaped bracts below it. The many cups are produced in umbels.

The six-inch- to one-foot-long stems are erect, and may be many-branched. A milky, latex-containing sap is present in the stem. Simple linear leaves lacking petioles are distributed alternately along the stem. Each is about one millimeter wide.

The genus name *Euphorbia* is in honor of Euphorbus, physician to an ancient king of Mauretania. Some spurge family relatives are the castor oil plant, the rubber tree, the cassava plant (which gives us tapioca).

Gerard wrote of it: "Some write by report of others, that it enflameth exceedingly, but my selfe speak by experience; for walking along the sea coast at Lee in Essex, with a Gentleman called Mr. Rich, dwelling in the same towne, I tooke but one drop of it into my mouth; which neverthelesse did so inflame and swell in my throte that I hardly escaped with my life. And in like case was the Gentleman, which caused us to take our horses, and poste for our lives unto the next farme house to drinke some milke to quench the extremities of our heat, which then ceased." A word to the wise . . .

The branching rootstock spreads cypress spurge far from the graveyard or garden where it was originally planted. Cypress spurge is, however, self-sterile, and thus if a colony springs up from but one plant by means of the spreading rootstock, seeds will not be produced.

At blooming time cut off the aerial stems and cut the rhizomes. They are at their most vulnerable at this time.

Cypress spurge [*Euphorbia cyparissias*]. **A.** Shoot with linear leaves, which when broken give out latex (a milky acrid juice). Leaves higher up are thinner. **B.** Umbellate cluster of flowers. **C.** Flower showing leaves under flowers, which are yellow. This flower is the female and the pistil can be seen. **D.** Mature fruit and two seeds.

D ×3   C ×3

F ×20

E ×

B

Yellow-Flowered Species

Portulaca oleracea L.

COMMON NAMES   Purslane, pusley, pursley, wild portulaca.
SOME FACTS   Annual; reproduces by seeds; introduced from Europe.
BLOOMS   July to September.
RANGE   Widespread throughout Canada and the United States.
HABITAT   Gardens, cultivated fields, waste places; in rich soil and dry soil.

Purslane, widely distributed in the Mediterranean region, was familiar even to Theophrastus, the ancient Greek known as the Father of Botany, and was listed by Dioscorides. Albertus Magnus, thirteenth-century general of the Dominican Order, also knew a great deal about plants and mentioned purslane in one of his books. It made its appearance in England in 1582.

It is a prostrate plant and forms many branches, giving rise to a mat. Its leaves are simple and obovate (wedge-shaped). There are thickened leaves, too, and these are filled with special tissue for the storage of water. Such leaves are said to be succulent. They are common on desert plants (called "xerophytes" by plant scientists), and purslane is really a desert plant, though it has invaded other places. The water-storage tissue sees it through drought.

The succulence is associated with a mucilaginous quality, which makes the leaves of this plant valuable as a thickener when added to soups and stews. The late Euell Gibbons offered, in his wonderful *Stalking the Wild Asparagus*, an interesting recipe for pickles using purslane leaves. They can be eaten cooked or raw. They make a good potherb, though purslane has never attained importance as a potherb in the United States as it has in India, Persia, and Europe for centuries.

From July to September small yellow sessile flowers open from flattened buds. The five yellow petals are inserted atop the sepals (which is uncommon among flowering plants) and open only in the sunshine. Following pollination, an urn-shaped, globular capsule results. It contains numerous seeds and opens by means of a round lid.

It was reported in Massachusetts in 1672. On October 7, 1749, Peter Kalm took the New Jersey ferry from Philadelphia, and while walking about noted, "The *portulaca* [sic] which we cultivate in our gardens [in Sweden] grows wild in great abundance in the loose soil among the corn. It was there creeping on the ground and its stems were pretty thick and succulent, which circumstances must give reason to wonder when it could get juice sufficiently to supply it in such dry ground."

If you really don't want it in your garden, then hand pull it.

Purslane [*Portulaca oleracea*]. **A.** Single prostrate plantlet. **B.** Alternate, simple, fleshy, wedge-shaped leaves. **C.** Five-parted flower. **D.** Many-seed globular capsule. **E.** Capsule opening. **F.** Tiny seed within.

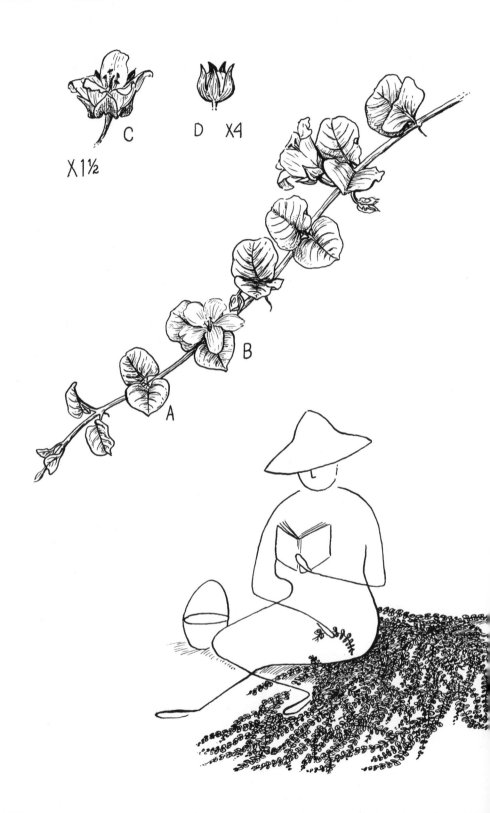

C

X 1½

D X4

B

A

Lysimachia nummularia L.

COMMON NAMES Moneywort, creeping loosestrife, herb twopence, two-penny grass, creeping jenny, creeping charlie, yellow myrtle.

SOME FACTS Introduced from Europe as an ornamental; perennial; reproduces from seeds and creeping stems.

BLOOMS June to July.

RANGE Widespread in the northeastern states.

HABITAT Fields, gardens, lawns, along ditches.

Moneywort's orbicular entire leaves look like coins and this feature has given the plant its specific name *nummularia* (from *nummulus*, Latin for coin). Several of its common names originate in the shape of the leaves. A related species was given the name *Lysimachia* in honor of King Lysimachus of Thrace by the great Dioscorides himself.

At one time creeping loosestrife was also known as *Serpentaria* because it was thought that wounded serpents could creep upon its stems and have their wounds mended. Herpetologists have been unable to verify this fact, however.

The smooth stem to which the round leaves are attached in opposite pairs creeps along the ground, rooting at each node when it gets the opportunity. Since it grows rapidly and branches frequently, creeping charlie can cover a good-sized area with speed, forming a matlike growth.

Pretty one-inch-wide wheel-shaped yellow flowers are found in the axils of the leaves. The five-parted corolla of yellow myrtle is spotted with small dark red dots.

Generally, this prostrate plant prefers damp ground, but you are just as likely to find it on a dry lawn or in a rock garden.

It has also been said of *Serpentaria* that it can stanch blood and heal human wounds as well as those of snakes. Gerard's suggestion for the use of *L. nummularia* sounds delicious, even if it is not very effective: "Boiled with wine and hony it cureth the wounds of the inward parts, and ulcers of the lungs; & in a word, there is not a better wound herb, no not Tabaco it selfe, nor any other whatsoever."

Dig out each rooted node. Better yet, catch the invasion early and uproot *Lysimachia* before it has formed a mat.

Moneywort [*Lysimachia nummularia*]. A. Flowering branch of this prostrate species showing the coin-shaped opposite leaves on short petioles. B. The five-parted corolla of a flower seen clearly. C. Flower from side showing sepals. D. Mature fruit (capsule), still protected by sepals.

A B C D E ♀ X4 F X2 X¾

Hieracium pratense Tausch.

COMMON NAMES Field hawkweed, king devil, yellow devil, yellow paint-brush.

SOME FACTS Introduced from Europe; perennial; reproduces by seeds and stolons.

BLOOMS June to early August.

RANGE Found very abundantly from Quebec to New York and Pennsylvania; less abundant to Michigan.

HABITAT Fields, meadows, roadsides, lawns, waste places.

Field hawkweed is a fairly common weed in fields, and, unfortunately, lawns. Except for the yellow heads it produces atop a slender, one- to two-foot-tall bristly stalk, it is very much like its close relative, devil's paintbrush or orange hawkweed (*Hieracium aurantiacum*), which has bright orange-red flowers. Both are formidable weeds.

The flower-tipped scapose stem—there are but two or three reduced leaves on it—arises from a basal rosette of narrowly oblong to lance-shaped leaves that taper back to margin-bearing petioles. They are bristly hairy on both surfaces. Small runners may be found on some plants.

This plant has been found to be sensitive to MCPA salt and 2,4-D amine and ester. Hunt for runners. If you leave any of them alive, the plant will return.

Field hawkweed [*Hieracium pratense*]. **A.** Rosette of leaves narrowly oblong to lanceolate. **B.** Stolon showing roots from lower surface and new plantlet developing. **C.** Small, hairy sessile leaves on flowering stalk. **D.** Loose cluster of heads atop stalk. **E.** A single flower of the head, which is composed of all ray flowers. **F.** A single head of now mature fruits (achenes), each with parachute (pappus).

H

G

F

E
X4

D

C

B

A
X1

Taraxacum officinale Weber.

COMMON NAMES Dandelion, lion's-tooth, blowball, cankerwort, milk witch, monk's-head, Irish daisy, priest's-crown, wet-the-bed.

SOME FACTS Introduced from Europe, originally from Asia; perennial; propagates by seeds and by forming shoots from the taproots.

BLOOMS March to December, but will bloom during warm winters.

RANGE Cosmopolitan weed.

HABITAT Fields, waste places, meadows, lawns.

Lion's tooth (dents-de-lion) with its sharp-pointed bright yellow "petals," which may, indeed, be very much like real lion's teeth in appearance, is the bane of the lawnkeeper. This is unfortunate, because it is basically a handsome, interesting, and useful plant.

Blowball has many values. If its rosette of leaves is taken up before flowering has begun, you will find it makes, after appropriate preparation, a very good potherb. It may even be added in its natural state to a salad. The taproot may be roasted in an oven, then ground and used precisely the way we use coffee. Best of all, the charming yellow flowers can be steeped in water and used to produce a delightful wine.

Each "flower" of the dandelion is actually composed of many individual flowers, and each "petal" is actually composed of five fused petals; so each "petal" indicates one entire flower. *Taraxacum officinale* is a member of the large taxonomic group known as the composites and is closely related to chicory (p. 223). Each flower is a ray flower.

Pick a "flower" of dandelion and pull it apart, spreading the pieces in the palm of one hand. Study each "piece" carefully—you will see that each is an entire flower!

The yellow flowering heads are produced from early spring to late, late fall (even into winter if the weather is warm). Each head, following seed set, becomes a ball of parachuted seeds; hence the common name blowball. These balls of easily detached parachutes have given pleasure to children for countless generations. Each bloom produces fifty or more parachutes, and each plant continues to produce heads of flowers throughout the growing season. Let one into your lawn and you are in trouble.

If you break the flowering stalk, you will note that it is hollow and that a milky substance appears at the site of the break. The milky juice contains latex, from which rubber could be made, though this is not commercially feasible at the present time.

Ridding yourself of *Taraxacum* can be accomplished mechanically by cutting off the tops before the flowers form and digging out the taproot. The plant is sensitive to 2,4-D amine.

Dandelion [*Taraxacum officinale*]. A. Single plant showing rosette (B) of pinnately lobed toothed leaves. C. Single head opening. D. Single quarter-sized head of yellow flowers. E. All the flowers in this head are ray flowers. F. Pollination accomplished, the head of mature, parachuted fruits. G. Single achene with parachute (modified sepals). H. Single dandelion plant showing the sturdy taproot.

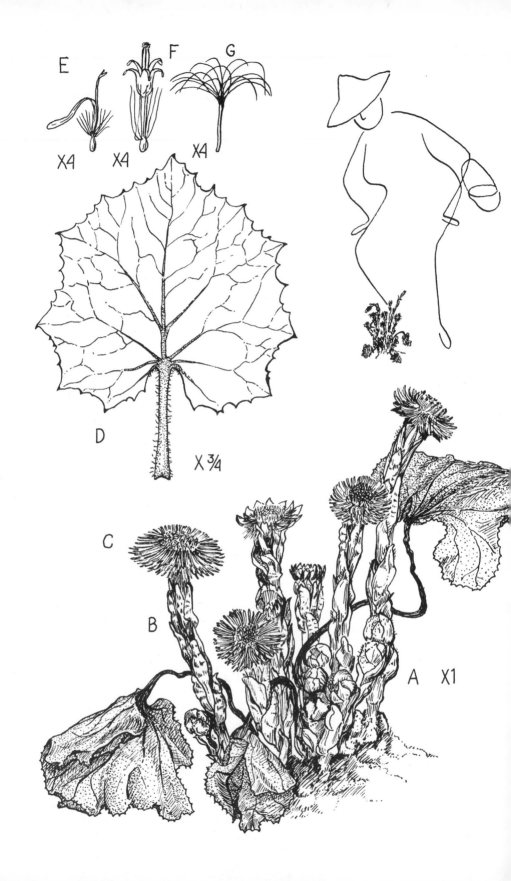

E   F   G

X4   X4   X4

D   X ¾

C

B

A   X1

Tussilago farfara L.

COMMON NAMES Coltsfoot, cough wort, ginger root, clay weed, horsehoof, foalfoot, dove dock.

SOME FACTS Introduced to United States from Europe; perennial; reproduces by seeds and rootstocks.

BLOOMS April to May.

RANGE Quebec to Minnesota and southward to Pennsylvania and Ohio.

HABITAT Moist clay soil, along stream banks, along brooks or dripping water, moist fields, roadsides.

The generic name comes from Latin words meaning "dispeller of coughs." The aid that coltsfoot or cough wort offers to the enrhumed has been known for thousands of years. Here's Dioscorides speaking of it in the first century A.D. (in Goodyer's seventeenth-century translation): "The leaves of this beaten small, & laid on doth cure ye Erysipelata, & all inflammation, but being suffited when it is dry, & ye smoke taken in through a tunnel [funnel], it cureth such as are troubled with a dry cough, & ye Orthopnea: when gaping they take the smoke in at ye mouth, & swallow it down."

John Gerard tells us, "A decoction made of the greene leaves and roots, or else a syrrup thereof, is good for the cough that proceedeth of a thin rheume. The green leaves of fole-foot pound with hony, do cure and heal inflammations. The fume of the dried leaves taken through a funnell or tunnell, burned upon coles, effectually helpeth those that are troubled with the shortness of breath, and fetch their winde thicke and often." It is still used to make a candy or cough drop, and Euell Gibbons in his delightful book *Stalking the Healthful Herbs* tells how to do this.

The common name coltsfoot arose because of the shape of the leaves produced by this plant at the end of its flowering season. Early in the spring slender, whitish scapes appear, growing from rhizomes. Each elongates to several inches high and is somewhat woolly-white, holding erect an inch-wide yellow single head of flowers, for which bees must, indeed, be grateful. These solitary, terminal heads are composed of both ray and disc flowers.

When flowering is all but over, the almost round leaves grow from the same rootstock on long petioles. Each is heart-shaped at its base, lobed and toothed, thick to the touch, and cloaked between with woolly-white hairs.

Since dove dock loves poorly drained soil, the easiest way to encourage it to move on is to restore good drainage. It is moderately sensitive to sodium chlorate and boric oxide or to a mixture of MCPA and 2,3,6-TBA salts. Keep in mind that the rootstock may be resistant.

Coltsfoot [*Tussilago farfara*]. **A.** Habit sketch of plant showing the coltsfoot-shaped leaves, some fading. **B.** The scaly-bracted flowering stalks, each with a single head. **C.** Single head of ray and disc flowers. **D.** Large coltsfoot-shaped leaf, woolly beneath and smooth above. **E.** Single ray flower. **F.** Single disc flower. **G.** Single fruit with parachute.

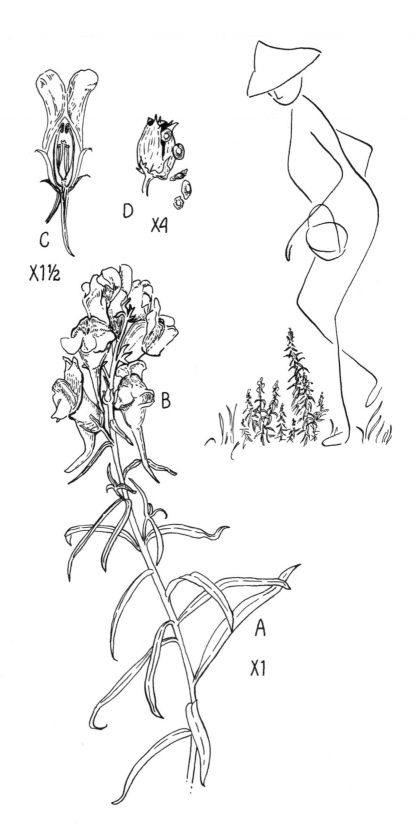

C

X1½

D   X4

B

A

X1

Linaria vulgaris L.

COMMON NAMES Butter-and-eggs, eggs-and-bacon, impudent lawyer, yellow toadflax, ramsted, flaxweed, wild snapdragon, Jacob's ladder.

SOME FACTS Introduced from Europe; perennial; reproduces by seeds and rootstocks.

BLOOMS May to September.

RANGE Eastern North America.

HABITAT Cultivated land, meadows and pastures, roadsides (often on gravelly or sandy soil).

Linaria vulgaris started out in America as a well-loved garden plant, the snapdragon of the hour. *Antirrhinum*, with its larger more varied-colored flowers, supplanted it, but not before it escaped to grace our lots and roadsides. Yellow toadflax is a relative of *Verbascum*, foxglove, Indian paintbrush, the monkey flower, and other snapdragons, all members of the snapdragon or figwort family.

The racemes of inch-long bright yellow flowers with orange throats are produced near the top of the stem. Each flower is composed of five sepals and five petals fused into a two-lipped yellow corolla with a long orange throat. These bilaterally symmetrical flowers are pollinated by bumblebees. Moths also get the nectar, but by stealing. Instead of landing on the lower lip, depressing it, and getting doused with pollen, a moth will insert its long proboscis through the lips and take nectar without becoming involved in the pollination of the flowers.

The wild snapdragon's stems are erect and the plant is found in colonies because the erect stems arise from an underground rhizome; individual stems rarely branch. Alternate, simple, linear leaves are present on the stems, attached without petioles. They are pale green (almost blue-green) and seem pointed at both ends.

The plant has never been reported to be poisonous, though it is apparently distasteful to cattle.

L. vulgaris is resistant to many chemicals such as 2,4-D or MCPA salt, but hot brine or caustic soda will kill plants. The underground rootstock must be cut as well or new plants will come up later.

Butter-and-eggs [*Linaria vulgaris*]. **A.** Flowering stem with linear leaves. **B.** Yellow and gold bilaterally symmetrical flowers. **C.** Single flower. If pinched, the upper portion separates from the lower as if a mouth is opening (hence, snapdragon). Drawn with "window" showing the stamens and pistil within. **D.** Mature fruit (capsule) opening and releasing the seeds. The power of magnification is denoted by X.

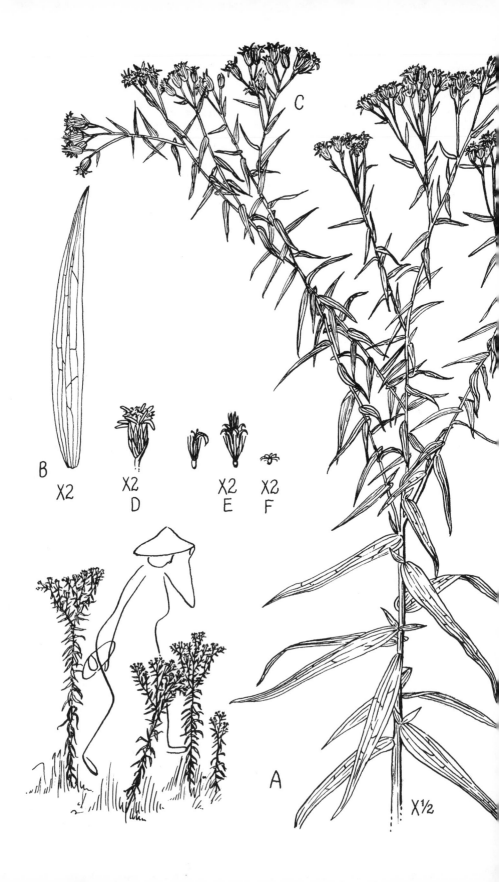

B
X2

D
X2

E
X2

F
X2

C

A

X½

Solidago graminifolia Salisb.

COMMON NAMES Narrow-leaved goldenrod, bushy goldenrod, creeping yellow weed, fragrant goldenrod.

SOME FACTS Native to North America; perennial; reproduces by seeds and rootstocks.

BLOOMS August and September.

RANGE Common in eastern North America, westward to Nebraska and Missouri.

HABITAT Open ground, thickets, roadsides, waste places.

The two- to four-foot-high grass-leaved goldenrod is very different from its close relatives, Canada goldenrod (*Solidago canadensis*) and the sea-side goldenrod (*S. sempervirens*). The narrow-leaved goldenrod has entire alternate lance-shaped to linear leaves, which are one to four inches long and one-fourth inch wide (or less). Each has three to five nerves and is sessile. It is, however, in the arrangement of the inflorescences that the difference is striking. The one-fourth-inch-long heads are clustered at the ends of a number of more or less erect branches, forming a somewhat flat- or dome-shaped inflorescence. Each plant has a number of these domes.

To destroy this plant, uproot while wearing gloves.

Narrow-leaved goldenrod [*Solidago graminifolia*]. **A.** Shoot showing thin leaves and floral branchings. **B.** A single linear leaf. **C.** Floral clusters (a composite). **D.** Single head of flowers. **E.** Single disc flower. **F.** Tiny seed with parachute.

B

C
X1

X1

A   X¾

Oenothera biennis L.

COMMON NAMES Evening primrose, field primrose, tree primrose, fever plant, night willowherb, wild beet.
SOME FACTS Native (but now introduced to Europe); biennial; seeds.
BLOOMS June to September.
RANGE Labrador to Florida and west to Rocky Mountains.
HABITAT Dry soil, fields, roadsides, waste places.

Long, long before the nineteenth-century botanist Hugo DeVries made the evening primrose famous as the plant on which he based his theory of mutation, simpler folk in Germany were using its robust, long taproot much the way we use parsnips (though only after cooking) and putting its rosette leaves in salads. The Greeks called the plant after the word for wine, *oinos*, and booty, *thera*, because they were convinced that eating the root increased one's capacity for wine. I have not found this to be so.

This relative of the lovely *Fuchsia*, and a member of the evening primrose family, first sends up from the top of its long, thick taproot a rosette of lance-shaped hirsute leaves some three to six inches long. During the second year a stem two to six feet in height, somewhat woody, emerges from the center of the rosette, and may branch from the base. Alternate leaves, which are narrow, lance-shaped, usually entire, and with short petioles, are found on the stem.

Bright yellow flowers that open during the evening hours make their appearance in the axils of the smaller leaves near the top of the stem. These sulphur yellow blooms are approximately one and a half inches in diameter and are generally closed during the day, opening very rapidly as the evening approaches; however, flowers on individual plants are open during the day. As the blossoms open, a fragrance becomes apparent and attracts the sphinx moths, which are so suggestive of hummingbirds. As the sun rises, the blossoms droop.

After fertilization the ovaries grow into inch-long capsules filled with seeds. These remain on the old dead stems long after they have dehisced at their summits. The seeds are very long lived.

While traveling in French Canada in 1749, Peter Kalm noted under the evening primrose, "An old Frenchman, who accompanied me as I was collecting its seeds, could not sufficiently praise its property of healing wounds. The leaves of the plant must be crushed and then laid on the wound."

If you are not in need of healing and do not want this fundamentally attractive weedy species, then dig out the first year's rosette, or cut the flowering stalk before flowering commences.

Evening primrose [*Oenothera biennis*]. A. Shoot showing hairy leaves and flowers. B. Sulphur yellow four-petalled flower that blooms at night (exceptions are abundant). C. The capsule, which opens along four lines of dehiscence.

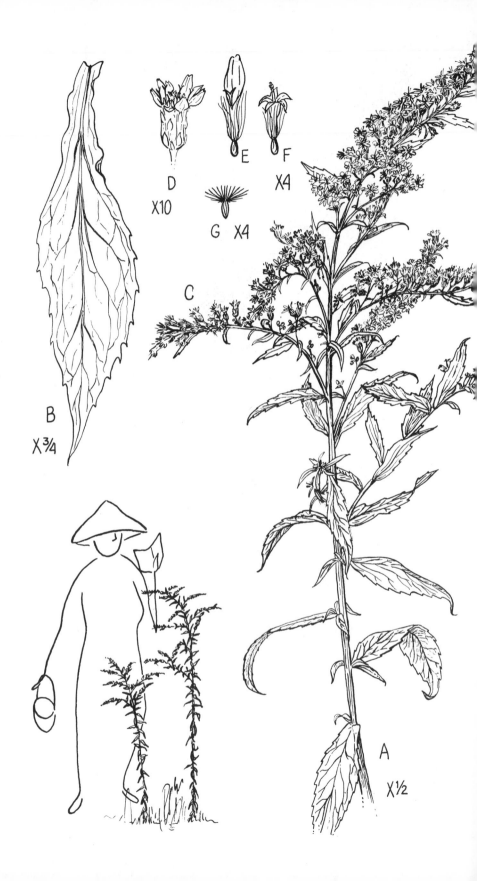

B
X¾

D
X10

E

F
X4

G X4

C

A
X½

Solidago canadensis L.

COMMON NAMES   Canada goldenrod, tall yellowweed, tall goldenrod.

SOME FACTS   Native to America; perennial; propagates by seeds and root-stocks.

BLOOMS   July to October.

RANGE   Widespread throughout eastern North America, occurs less frequently westward to British Columbia.

HABITAT   Rich, open soil, thickets, meadows, waste places.

Solidus, meaning "to make whole" in Latin, refers here to the fact that certain members of the very large and taxonomically confusing genus *Solidago* were used medicinally. Listen to John Gerard writing in the late sixteenth century: "It is extolled above all other herbes for the stopping of bloud in bleeding wounds; an hath in times past beene had in great extimation and regard than in these daies; for in my remembrance I have known the dry herbe which came from beyond the sea sold in Bucklersbury in London for halfe a crowne an ounce."

Goldenrod is a very abused plant. It is hotly hated by hayfever sufferers. This is sad, for they have no just reason for hating goldenrod. It merely produces its gorgeous golden blossoms at the very moment hayfever sufferers happen to begin suffering. The two events, however, are not related. Rather it is the wind-borne pollen of *Ambrosia trifida* (p. 117) or *A. artemisiifolia* (p. 119) that produces the largest share of the suffering. Since the pollen of goldenrod is not wind-carried but insect-carried, the only way your nose will be disturbed by goldenrod pollen is if an insect that recently visited a blossom of goldenrod inadvertently enters your nose. Because of their good sense, the people of Alabama, Kentucky, and Nebraska have taken this lovely American flower as their state flower despite its undeserved evil reputation.

Under excellent growing conditions, specimens of Canada goldenrod that are eight feet high can be found, but one is more likely to encounter plants that are three to six feet tall. Canada goldenrod has alternate leaves that are narrowly lanceolate and essentially uniform from their base to their summit. Each is serrate, three-nerved, smooth along the upper surface, and hairy beneath. Leaves found lower down on the stem are petiolate, while those found higher up are usually sessile and entire.

The heads of this composite are borne along the upper surface of several flowering branches that curve outward and often downward. The individual heads are small and dull yellow, each with ten to twenty ray florets.

If you wish to remove the plant, hand pull (be sure you have your gloves on).

Canada goldenrod [*Solidago canadensis*]. A. Shoot showing leaves and flowers. B. Single leaf with flared petiole. C. Floral branches. D. Single head of flowers. E. Single ray flower. F. Single disc flower. G. Single seed with attached parachute (modified sepals).

A   X½

C

B

X¾

X2 D   X2 E   X2 F   X2 G

Solidago sempervirens L.

COMMON NAME Seaside goldenrod.

SOME FACTS Native to the United States; perennial; reproduces by seeds and rootstocks.

BLOOMS August to October.

RANGE Along coast from Gulf of St. Lawrence to Florida and Texas.

HABITAT Saline places, brackish water, rear of beaches.

The seaside goldenrod will probably never bother you as a weed unless you are fortunate enough to own some shoreline. Even then, this fundamentally attractive shore plant will very likely not disturb you. It lives in wet, usually salty places near the coast, and often forms a border approximately where high tide makes a line along the beach or shore.

The taxonomy of the genus *Solidago* is difficult, and even the experts may differ as to which species is which in some cases.

It has thick, succulent leaves, which are often noted on seaside or salt-loving species. The basal leaves are long-stalked. Like *Solidago canadensis* (p. 145) but unlike *S. graminifolia* (p. 141), its heads of flowers are along the upper side of branches that curve outward.

Seaside goldenrod [*Solidago sempervirens*]. A. Shoot showing distribution of heads and leaves. B. Single succulent leaf, which is thick to the touch. C. Floral heads. D. Single head. E. Single disc flower. F. Single ray flower. G. Single seed (achene) with attached parachute. The power of magnification is denoted by X.

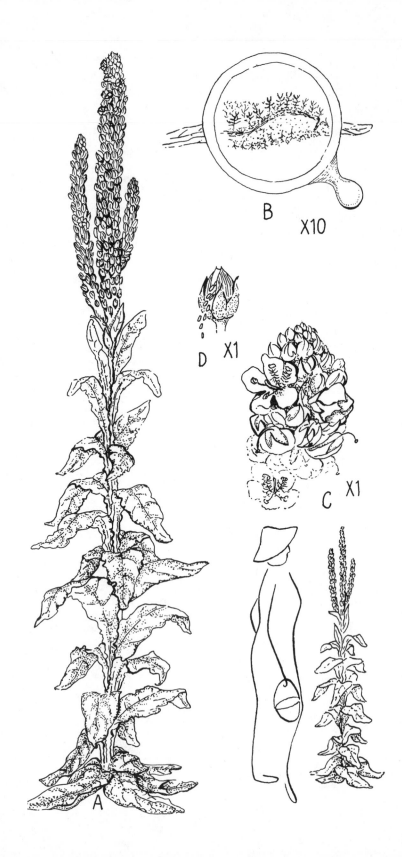

B X10

D X1

C X1

A

Verbascum thapsus L.

COMMON NAMES Mullein, flannel leaf, velvet leaf, velvet dock, hedge taper, candlewicks, blanketleaf, Aaron's rod, great mullein.

SOME FACTS Introduced from Eurasia; biennial; reproduces by seeds.

BLOOMS July to September.

RANGE Abundantly found in north temperate America.

HABITAT Field, roadsides, waste places, pastures; likes dry, stony, or gravelly soils.

There are few weeds with so many common names, and few that are as interesting as Verbascum thapsus. Even its scientific name is interesting: Verbascum is a misspelling of barbascum, the Roman name for the plant, which means bearded, and was therefore quite appropriate for this plant since it is covered with a dense coat of hair. Its specific name comes from the city of Thapsus, an ancient town in North Africa (now Tunisia) where in 46 B.C. Julius Caesar defeated Pompey.

Velvet dock, felt wort, blanketleaf, flannel leaf—all these names refer to the hairs covering the large leaves on both the rosette and the stalk. This plant's most common name, mullein, also refers to the coating of hairs, for mullein comes from an old English word for soft (old French moll, soft).

A rosette of fuzzy leaves ten to twenty inches long and four to six inches wide appears during the first year of this biennial's growth. The leaves, because of the downy coat, appear white to gray-green.

During the second year's growth a stalk two to seven feet high grows from the center of the rosette, and may branch from one to two times near its top. The leaves that are found on the flowering stalk become progressively smaller from the bottom to the top of the flowering stalk. All of them, however, are densely hairy.

Yellow flowers open all along the stalk near the top and are nearly sessile, with very woolly sepals and sulphur yellow petals. The stamen hairs give one pollinator, the hover fly, a good grip as she lands. A globular capsule develops one-fourth inch long, each of whose two chambers contains many tiny brown seeds.

American Indians smoked its dried leaves. (Awful when I tried it!) Roman ladies dyed their hair yellow, not with peroxide, but with water in which the yellow petals of mullein had been steeped. Quaker ladies rubbed the bristly young leaves against their rouge-forbidden cheeks.

Dig out the rosette as it appears, and if you have waited until the second year's flowering stalk appears, destroy it before seed set.

Mullein [Verbascum thapsus]. A. Whole plant showing first year's rosette of leaves and (at top) second year's branched flowering stalk. B. Leaves are covered thickly with branched hairs. C. Top of a flowering stalk showing the many flowers, each with yellow petals and woolly stamens. D. Single capsule shedding tiny seeds.

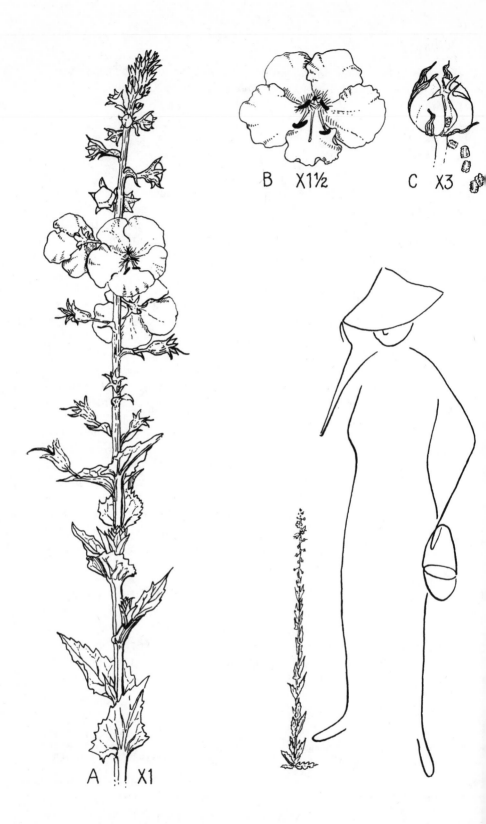

A X1

B X1½

C X3

Verbascum blattaria L.

COMMON NAME Moth mullein.
SOME FACTS Introduced from Europe; biennial; reproduces by seeds.
BLOOMS June to November.
RANGE Quebec to Minnesota and south to Florida and Kansas, west to the Pacific and south to the Gulf of Mexico.
HABITAT Meadows, pastures, waste places.

That the moth mullein has attracted far less attention than its cousin, the great mullein (p. 149), is attested to by its failure to have more than one common name. Two opinions exist as to why this weedy plant has come to be called *moth* mullein. One is that the night-flying moths pollinate the blooms, and the other is that the open flower, with its purple-haired stamens, looks like a moth. Opinion seems to be equally divided, too, on what the specific name *blattaria* refers to. It has been suggested that the leaves repel the cockroach (*blatta*), for which there is not a shred of evidence.

Moth mullein, unlike its cousin, is not covered with hairs, though a few glandular hairs may be found near the top of its two- to five-foot stem. The stem of this biennial begins at the center of the rosette and rises round, slender, and usually unbranched. The petiolate lower leaves are oblong, though sometimes they may be pinnatifid, hairless, veiny, dark green, and anywhere from three inches to one foot long. Leaves higher on the stem are sessile and pointed.

Both yellow- and white-flowered specimens are known, though apparently the yellow-flowered type is more common. The upper petal is usually found to be brown on its back, while the filaments of the unequally sized anthers are coated with purple hairs. Flowering begins at the bottom of the open raceme with three- or four-inch-wide blossoms appearing at one time. The upper flowers are still opening after the lower blossoms have produced their globose, many-seeded capsules.

Moth mullein might easily be included among the more respectable plants of the cultivated garden but for the fact that so few blossoms open at any one time. Were all the flowers to open at once, the flowering stalk would indeed be very attractive.

If the rosettes are dug out soon after their appearance, moth mullein should give you no further trouble.

Moth mullein [*Verbascum blattaria*]. A. Shoot showing nonhairy leaves of this mullein. B. The five-petalled white or yellow flower. C. Globose capsule shedding a few seeds. The power of magnification is denoted by X.

C

D X1   E X2   F X2

B

A

X 3/4

Helianthus annuus L.

COMMON NAMES Sunflower, wild sunflower.
SOME FACTS Native to the United States; annual; reproduces by seeds.
BLOOMS July to September.
RANGE Frequently seen east of the midwestern states, but native to region between Minnesota and Saskatchewan, and southward to Missouri and Texas.
HABITAT Meadows, fence rows, roadsides, waste places.

Who has not marveled at the huge, sometimes foot-wide flowering heads of the cultivated sunflower, native to our western prairies? Kansas boasts the sunflower as its state flower, and many of us can still recall its use as a symbol in a presidential campaign.

The noncultivated form stands one to six feet high and the cultivated forms are up to fifteen feet tall. Broad, oval, pointed, cordate, three-ribbed leaves that are three inches to a foot long emerge from the tall, rough stem on stout, hairy petioles. Flower heads are three to six inches wide, with yellow, sterile ray florets, and the disc florets are five-lobed, tubular, dark brown, and fertile. Its well-known seeds (achenes) are ovate to sedge-shaped, flat, and white, gray, or dark brown with lighter stripes, or are gray-mottled.

The American Indians used the seeds for food and oil. They were parched and ground into flour, made into bread or cakes, or put into soup. The shells, roasted or crushed, are said to make a drink that approximates coffee in taste. Kalm reports that the Indians of French Canada mixed the seeds into sagamite or corn soup. In Russia sunflower seeds are eaten as we eat peanuts. Parrots love them, too.

Use your garden gloves when uprooting this very American plant because it is rough to the touch.

Sunflower [*Helianthus annuus*]. **A.** Portion of flowering stalk. **B.** Large, heart-shaped, rough, hairy leaf. **C.** The head of ray and disc flowers. **D.** Single ray flower. **E.** Single disc flower. **F.** Seed.

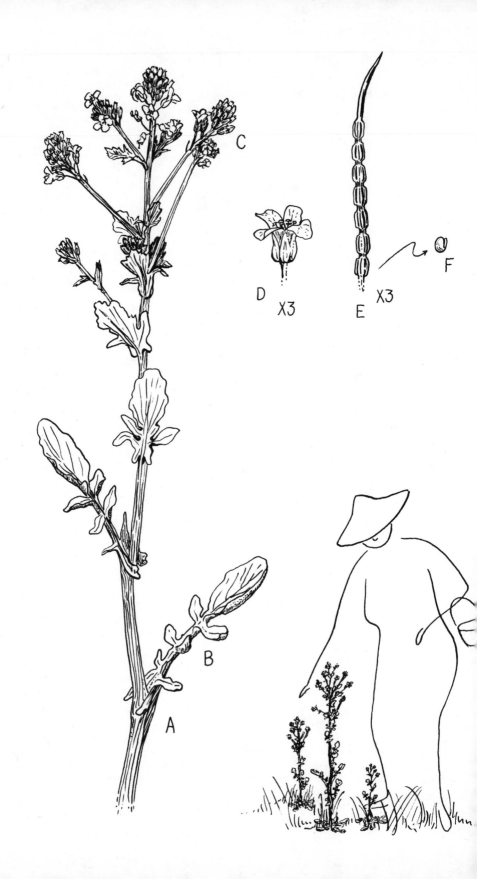

C

D X3

E X3

F

A

B

Barbarea vulgaris R. Br.

COMMON NAMES Yellow rocket, winter cress, St. Barbar's cress, bitter cress, herb barbara, rocket cress, water mustard, potherb.

SOME FACTS Introduced to the United States a second time by colonists, who found it already here; weakly perennial (really biennial); reproduces by seeds.

BLOOMS May to June.

RANGE Abundant in northeastern and north central states, occasionally in Pacific Northwest.

HABITAT Fields, meadows, roadsides, waste places.

Though the generic name was given in honor of St. Barbara (as were several of the common names), patroness of canoneers and miners and protectoress of those caught in thunderstorms, there is nothing about winter cress that can be directly compared with anything about Barbara. The name arises from the fact that the young leaf rosettes are available even on December 4, St. Barbara's Day, though you may have to dig through snow to find them.

The green-in-winter rosette leaves, which are above the fibrous root system, and the leaves lower on the flowering stalk are pinnatifid, the terminal lobe being larger than the others. Upper-stem leaves are coarsely toothed and clasping. All the leaves of herb barbara are dark green, smooth, and shiny.

Branches of the one- to two-foot-long stem end in racemes of bright yellow four-petalled flowers, the petals forming the Maltese cross pattern usual among members of the Crucifers (p. 203). The flowers are sweetly scented.

Both leaves and unopened flowers are useful. Winter cress leaves are used in salads, though they should be collected before late March if they are not to be too bitter. These leaves also make an excellent potherb. The unopened flower buds may be cooked exactly as are our vegetable that is composed of unopened flower buds, cauliflower. The leaves, which contain a high amount of vitamin C, are favored by sheep and cattle.

Inch-long pods (siliques) are seen not long after fertilization of the flowers. These are nearly erect and almost four-angled.

Using your garden gloves, pull the plant up as it begins to flower. If you hoe out the rosettes, put a little dry salt over the stump hole and you'll rid yourself of yellow rocket.

Yellow rocket [Barbarea vulgaris]. A. Flowering stalk arising from overwintering rosette. B. Single leaf showing margins cut into sections and lobed at the base. C. Small racemes of four-petalled flowers. D. Single four-petalled flower. E. Valved pod (silique). F. A single seed.

C

D

X4

E

F

X4

Ranunculus acris L.

COMMON NAMES   Tall field buttercup, tall crowfoot, butter flower, meadow buttercup, blister plant, goldcup, kingcup.

SOME FACTS   Introduced from Europe; perennial; reproduces by seeds.

BLOOMS   May to September.

RANGE   Throughout the United States and Canada; northwest Washington.

HABITAT   Pastures, meadows, roadsides, waste places.

Thinking of the buttercup as a weed may seem strange to some, but in a poorly drained meadow or sections of some lawns, tall crowfoot can become a pest. Yet, we must admit that goldcup, weedy or not, is a charming plant. However, let your admiration remain ocular because blister flower produces an acrid (hence *acris*) juice that blisters the mouths and intestines of cattle and will blister your skin.

The inner surface of the cup of goldcup seems to glisten in sunlight. This is due to the presence of many starch grains in the cells of the epidermal (surface) layer of the petals.

Since tall crowfoot will not do well in well-drained soil, there lies a way out if this plant has become a pest. If you plan to hand pull, wear leather gloves as the plant's juice may blister your skin.

Tall field buttercup [*Ranunculus acris*] **A.** Flowering stalks showing deeply palmately incised leaves. **B.** Single palmately trisected and deeply dissected leaf. **C.** Flower showing radial symmetry. **D.** The five petals and the numerous stamens are seen here. **E.** The cluster of fruits (follicles) showing that there were numerous pistils. **F.** Single seeds (achenes).

E

F   X6

D   X2

B

C

X¾   A

Lotus corniculatus L.

COMMON NAMES Bird's-foot trefoil, cat's-clover, crow-toes, ground honey-suckle, sheep-foot, hop o'my thumb, devil's-claw.
SOME FACTS Introduced from Europe; perennial; reproduces by seeds.
BLOOMS June to August.
RANGE Widespread in the United States.
HABITAT Waste places, lawns, roadsides.

Bird's-foot trefoil, sporting its dark green foliage and bright yellow, typically papilionaceous blossoms, is a lovely thing to behold. In the garden, however, its spreading habit of growth causes it to grab space in which other plants might grow. It escaped from cultivation long ago and is now a troublesome, albeit attractive, weed. In Europe, whence it hails, it is not only a troublesome weed, but is also used as a popular forage plant despite the fact that it is known to contain potentially toxic amounts of a cyanogenic substances. Crow-toes is also grown in eastern New York, western Oregon, and northern California as pasture, and is mixed with other plants to make a good hay.

Hop o'my thumb, a winter hardy plant, has a stem that may be spreading, but if a number of plants are growing closely packed, the stems may be ascending or even erect. Under good growth conditions, they may reach a length of twenty inches. Sessile, alternate compound leaves are borne on the stems and each is composed of five linear to oval leaflets, the two lower of which are right next to the stem and resemble stipules. The bright yellow flowers occur in umbel-like stalked clusters in the axils of the upper leaves. Interestingly, the flowers may be found in shades from bright yellow to brick red. Following pollination, the pods cluster at the tip of the pedicel, and when mature, look like a bird's foot, hence the common name.

If bird's-foot trefoil appears on your lawn, dig out the plants before they seed. It is moderately resistant to MCPA salt and 2,4-D amine, which means that repeated efforts will probably succeed.

Bird's-foot trefoil [*Lotus corniculatus*]. **A.** Habit sketch of whole plant showing (**B**) the pinnately-compound leaves (five leaflets). **C.** The bright yellow papilionaceous flowers. **D.** One flower somewhat enlarged. **E.** A cluster of mature fruits (legumes). **F.** A single, much enlarged seed. The power of magnification is denoted by X.

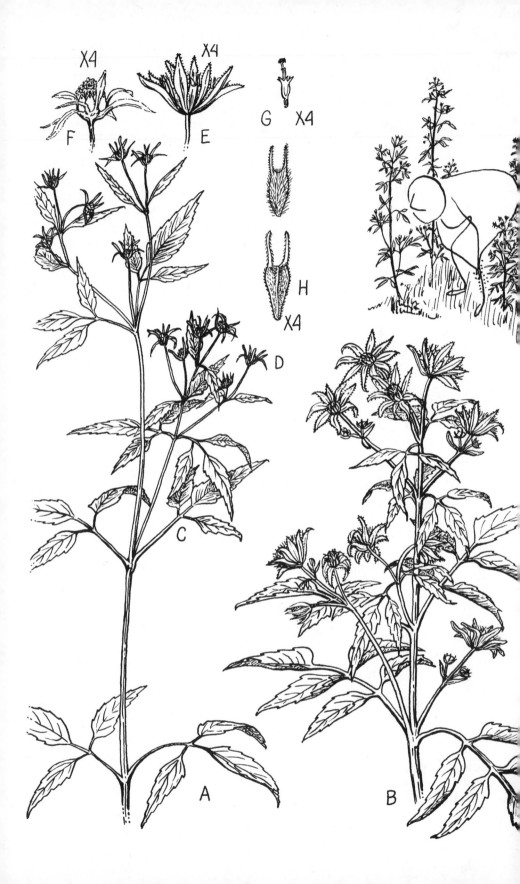

Bidens frondosa L.

COMMON NAMES Beggar ticks, stick-tights, devil's bootjack, bur-marigold, pitchfork-weed, tickseed sunflower.

SOME FACTS Native to the United States; annual; propagates by seeds.

BLOOMS July to October.

RANGE Widespread throughout the United States and into southern Canada.

HABITAT Moist soil, especially in gardens, roadsides, pastures, waste places.

Beggar ticks forces you to pay it heed. After a nature walk or field trip (a stroll through an unattended, poorly drained garden may do), it may take you a very long time to remove all the stick-tights stuck tightly to your clothes, sometimes in very great numbers.

Beggar ticks is merely trying to spread its seeds far and wide, and when its fruits (containing a single seed) become stuck to animals' fur, they are carried far before being dropped.

Bidens (which means having two teeth) grows just about everywhere, though it is particularly partial to poorly drained places. Its freely branching shoots may be five feet high, though some are but two feet high and take on a purplish hue.

The three- to five-parted compound leaves are oppositely disposed along the stems; just a short distance below the terminal blossoms are found simple, lanceolate, toothed leaves, but a bit further down three-parted compound leaves are found, and still further down, five-parted compound leaves. Each leaflet is lanceolate and sharply toothed. The central leaflet of compound leaves tends to be on a longer petiole than leaflets on the sides. All are smooth.

Beggar ticks is a composite, and its head of flowers is composed of both ray and disc flowers, though sometimes the ray flowers (always very small) are missing altogether. Both are yellow-orange in color. Only the disc flower produces the seed in this plant.

The fruit is a flat achene, black in color and wedge-shaped; it is ridged down the center of each face. Two diverging points are at the apex; each is covered with downward-pointing barbs. It is the fruit that makes the stick-tight so unpopular, though it has also been reported that the milky juices in the stems and leaves produce itchy skin in some people.

Pull the plants out before seed sets, or better yet, since it tends to favor poorly drained soil, improve the drainage where you find it growing and this may help you get rid of it.

Beggar ticks [*Bidens frondosa*]. **A.** Portion of the flowering stem showing the pinnately-compound leaves (three leaflets). **B.** Portion of the stem of *B. vulgata* with its five-parted pinnately-compound leaf. **C.** Single pinnately-compound leaf (three leaflets) of *B. frondosa*. **D.** Small head of flowers. **E.** One head enlarged to show ray flowers. **F.** Some of the ray flowers pulled away to reveal the disc flowers at the center. **G.** Single ray flowers. **H.** Two-toothed (bidens) fruit (achene), which sticks to fur or clothing.

X3 D

E

X2

C

A X¾

B

X¾

Pastinaca sativa L.

COMMON NAMES Wild parsnip, field parsnip, madnip, bird's-nest, hart's-eye, tank, siser.

SOME FACTS Introduced to United States from Europe for its roots and escaped early; biennial; propagates by seed.

BLOOMS June to August.

RANGE Widespread in the United States and Canada.

HABITAT Waste places, roadsides, meadows; on rich soil and usually where soil is moist.

Wild parsnip earned its generic name because its roots are dug up (*pastinare* in Latin means to dig), though some etymologists believe the name originates in the Latin *pastus*, food. Its wild, edible roots were consumed in ancient Greece and Rome. Dioscorides wrote of it, "Siser is commonly knowne, whose root being sod is pleasing to ye tast & profitable for ye stomach, vreticall & stirring up of the appetite."

A carrot relative, wild parsnip is also a biennial. During the first year there arise from the long taproot long pinnately-compound leaves that are composed of coarsely toothed and lobed segments. These are attached to the top of the taproot by long, flattened, and grooved petioles. During the second season of growth there arises from the center of the rosette a two- to four-foot-tall hollow and deeply grooved light green stem whose leaves are much smaller (some are clasping). Flat-topped compound umbels composed of small yellow blossoms, each of which gives rise to a mericarp, cap the stalk.

The large, famous taproot grows deep within the soil and is composed mainly of phloem tissue filled with starch. After this root has been exposed to cold, it develops a sweet taste.

The American field parsnip is said to be poisonous during its second year, while its European counterpart is thought not to be. It is reported that some people are sensitive even to the touch of the leaves and soon develop a rash.

Since wild parsnip serves as host to a fungus that also attacks the related celery plant, it is necessary to eradicate *Pastinaca* from fields near celery-growing regions. As a weed in nonfarm areas, it will usually be found under moist conditions. If three to five pounds per acre of 2,4-D in eighty gallons of water are applied between March and May or August and October, the plants will be destroyed. If you choose to pull them out, wear gloves.

Wild parsnip [*Pastinaca sativa*]. **A.** Pinnately-compound leaf of first year's rosette. A very large, grooved leaf, each leaflet of which is deeply toothed and often cut-lobed. **B.** The flowering stalk that arises the second year, bearing smaller pinnately-compound leaves. This stalk is grooved and hollow. **C.** Large flat umbels. **D.** Single flower from an umbel. **E.** The fruit (a mericarp). The power of magnification is denoted by X.

E X6

X9 D

C

B

A

Tanacetum vulgare L.

COMMON NAMES Tansy, bitter buttons, ginger plant, parsley fern, hind-head, bitter weed.

SOME FACTS Introduced by colonists; perennial; reproduces by seeds and rootstocks.

BLOOMS July to September.

RANGE Widespread in the United States, especially in the Northeast; less frequently seen in the West.

HABITAT Roadsides, waste places, old yards on dry, sandy, or gravelly soils.

Tansy oil, distilled from the leaves and tops of this delicately leaved plant, has been used in medicine since the Middle Ages. It has long served as a home remedy for ridding children of worms and also to bring about abortions, though injudicious use of bitter buttons for this purpose has caused loss of life. A slight overdose will produce severe gastritis or convulsions. The oil was also mixed with that of fleabane and pennyroyal, diluted with alcohol, and used as a mosquito repellent before the advent of the many commercial insect repellents. Even after death, the plant was thought to be of some help; it was placed within the winding sheets to inhibit the activity of worms.

The stem of this early escape from colonial cultivation stands one to three feet high and generally remains unbranched. It is the flowering part of the plant, at the top, that branches many times. Tansy's delicate leaves are beautiful. Each is deep green, smooth, and one- to three-pinnately divided. Leaf segments are narrow and toothed. Leaves higher on the stem are smaller and less divided than those lower down.

The common name bitter buttons refers to the button-shaped, clustered flowering heads (tansy is a composite). Each button is one-half inch across, yellow, and composed entirely of disc flowers. The central flowers are perfect, while the marginal ones are usually pistillate.

In England tansy is given the generic name *Chrysanthemum*, which is probably more accurate, but no matter which generic name you use, you will find the plant moderately resistant to 2,4-D amine or MCPA salt. Pulling the plant up and then spraying the area with sodium chlorate should kill the rootstocks.

Tansy [*Tanacetum vulgare*]. A. Single flowering stalk. B. A pinnately divided (sometimes three times) leaf with winged petioles. C. Terminal corymbose clusters of buttonlike heads of flowers, of which many are pistillate. D. Single disc flower with both stamens and pistils. E. A single disc flower that is pistillate.

Potentilla canadensis L. (*Potentilla recta*)

COMMON NAMES Common cinquefoil, five-finger, bareen strawberry.
SOME FACTS Native to the United States; perennial; propagates by seeds and stolons (runners).
BLOOMS June to September.
RANGE Common from New England to the Great Lakes and southward.
HABITAT Dry soil; fields, meadows, pastures, waste places.

John Gerard, the herbalist, noted that following names for our cinquefoils: *Quinquefolium* (Latin); *Pentaphyllon* (Greek); *Cinquefoglio* (Italian); *Quintefueille* (French); *Cinco en rame* (Spanish); cink-foile, five-finger, and five-leaved grasse, sinkfield (English). It does *not* have five leaves but three, and they are not leaves at all but leaflets! The lateral two are so strongly parted they look like two separate leaves.

Five-finger has very slender, hairy stems that are spreading and procumbent and may be two feet long. Stem tips touching the earth give rise to new plantlets. The five-petalled flowers, yellow and one-half inch wide, are within the leaf axils of the first two nodes. Many stamens and pistils may be found around a plump central receptacle.

Do not confuse this species with *P. argentea*, also very common and similar in appearance and growth. *Argentea* means silvery, and the leaves of this species have lower surfaces that are silvery green; each leaflet is more deeply serrated. Nor could you confuse the common cinquefoil with the sulphur cinquefoil, *P. recta* L., an introduced species which is perennial, erect, and may grow two feet tall. This species is strikingly robust and woody; its leaves are quite different from the other species, and its sulphur yellow flowers are often clustered in cymes. It is not rarely used as a garden plant, but it must be kept under control or it will take over because of its persistent root.

Peter Kalm reports of the common cinquefoil that a liquid made from boiling its leaves in water was thought to be a good remedy for fevers and ague and that "several persons recovered by this means."

Species of *Potentilla* are resistant to most chemical weed killers but are moderately sensitive to NaCl and boric oxide. A few treatments of sulphur cinquefoil with MCPA salt and 2,4-D amine and ester may be effective. The presence of cinquefoil indicates an impoverished soil; if you enrich it you may not see these plants again.

Common cinquefoil [*Potentilla canadensis*] and sulfur cinquefoil [*Potentilla recta*]. **A.** Branch showing general habit. **B.** Flat-growing branch; more erect posture of *P. recta* (a), flatter posture of *P. canadensis* (b). **C.** Single compound leaf of three leaflets, two at side parted to give appearance of five leaflets. **D.** Five-parted flower, bright yellow. **E.** Opening fruit showing escaping achenes.

D E F G

C

B

A

Rudbeckia hirta L.

COMMON NAMES Black-eyed Susan, brown-eyed Susan, coneflower, ox-eye
daisy, yellow daisy.
SOME FACTS Native to our shores; biennial; reproduces by seeds.
BLOOMS July to October.
RANGE Common in East, often seen in Midwest and West.
HABITAT Prairies, meadows, pastures, waste places, old fields.

It is difficult to dislike the lovely black-eyed Susan, whose generic name
is modeled after Linnaeus' great teacher, Dr. Rudbeck, but the truth is
that once it invades a field, it spreads very rapidly. Ox-eye daisy hails from
the plains and prairies of the West and Midwest; no doubt it was a
stowaway on trains coming East.

During its first year a rosette of simple, two- to six-inch-long, spatulate,
three-nerved leaves forms, which are rough, hairy, and have grooved
petioles. An erect stem, one to three feet high, which is simple or sparingly
branched near its base, arises from the center of the first year's rosette.
The stem and the simple, oblong, sessile stem leaves are also rough and
hairy. Two- to four-inch-wide heads occur singly. A rounded brownish-
black cone of disc florets is seen at the center of the flower and is surrounded by fourteen to sixteen long, sterile ray flowers that are bright
orange-yellow in color.

Coneflower has been responsible for poisoning sheep and hogs in
Indiana. If it appears in your lawn or field consistently, remove the basal
rosette leaves and the plant should perish.

Black-eyed Susan [*Rudbeckia hirta*]. **A.** Flowering stalk with the simple, somewhat
hairy leaves. **B.** Large head showing ray and disc flowers. **C.** Head of pollinated flowers.
D. Single ray flower. **E.** Single disc flower. **F.** One hairy stigma enlarged. **G.** Single fruit.

X2 C

D
X2

E
X2

B

A
X1

Impatiens biflora Willd.

COMMON NAMES Jewelweed, touch-me-not, snapweed, wild lady's slipper, silver cap, wild balsam, lady's eardrop.
SOME FACTS Native; annual; reproduces by seeds.
BLOOMS June to September.
RANGE Newfoundland to South Carolina, Arkansas, Alabama, Oklahoma.
HABITAT Moist woods, brooksides, near springs.

Touch-me-not's inch-long seed pods (capsules) live up to this common name when mature. The faintest disturbance and they explode into five shreds, hurling their seeds far and wide and even making a small but not inaudible noise.

Jewelweed's glabrous (smooth) branching stems are almost clear, something that may be difficult to observe in the deep shade of the stream banks where it lives and grows to a height of two to four feet. Young shoots may be cooked and eaten, though some say touch-me-not is poisonous; however, Professor John Kingsbury of Cornell University does not list it among the poisonous plants in his book *Poisonous Plants of the United States.*

The unwettable, thin, one- to three-inch leaves of touch-me-not are pale green and ovate to elliptical in shape. They are blunt-ended, with crenately toothed margins, and are attached on longish petioles.

The lovely orange-yellow flowers occurring in twos from slender pendent stalks are mottled reddish-brown, and are spurred, but the spur is on a petal-like sepal that ends in a sac contracting into a long in-curved spur.

Both flower and leaves are used to dye wool yellow, as Peter Kalm tells us. He calls the plant "the crowing cock," no doubt thinking of its horizontal flowers on their nodding stalks. Indians used it to sooth itches, unhealthy scalp, athlete's foot, and dermatitises of a fungal nature, but the most common use to which this plant has been put for centuries is as an antidote to the poison of poison ivy.

The juicy stems rubbed against an area recently in contact with the shiny, poisonous leaves of *Rhus radicans* (p. 61) will prevent the appearance of the dreaded blisters, and if they have already appeared, will help rid you of them with speed. It is said that if one bathes in a tub into which juices of *Impatiens biflora* have been poured, one will not be affected by poison ivy. Several drugstore anti-poison ivy lotions and remedies contain the juices of *Impatiens.*

Pull up these annual plants before flowering, but remember, it might not be a bad idea to have a patch just in case poison ivy comes your way.

Jewelweed [*Impatiens biflora*]. A. Portion of a branch showing the thin leaves, which are ovate to elliptical. B. One of the two blossoms produced together on small pendent stalks. C. Single flowers showing bilateral symmetry. D. Single fruit (capsule). E. Same fruit, merely touched, explodes and throws seeds out of capsule.

C
X3

D
X3

E
X9

B

A

X3/4

White-Flowered Species

Cuscuta gronovii L.

COMMON NAMES Dodder, gold-thread vine, onion dodder.

SOME FACTS Native; reproduces by seeds and small pieces of overwintering stem; fundamentally an annual.

BLOOMS July to September.

RANGE Widespread in northeastern United States.

HABITAT Low wet fields, marshy thickets, waste places.

Few flowering plants are parasitic, and those that are seem to occur within certain orders of flowering plants, especially among snapdragon relatives or morning glory relatives. *Cuscuta*, as well as many other angiosperm parasites, effects a connection with the host plant by means of its roots, which dig deep into the tissues of the stems of the host, stealing water, minerals, and foods. Though dodder has been shown to have a small amount of chlorophyll, and is, therefore, capable of some photosynthesis, it does not carry on enough photosynthesis to permit manufacture of all the food it requires.

Dodder will attach itself to just about any flowering plant, and does so in forty-seven of our fifty states. Following germination from a seed in the soil, the stem, on reaching a neighboring plant of a different species, will send its haustoria (roots) down into the stem tissues of the host. In one test dodder produced 2,405 feet of stem from a single seed germination in a single growing season of four months! The longest single stem was six feet nine inches.

No leaves cloak dodder's yellowish stem, but small scales take their place. The yellow is due to the abundance of yellowish and orange carotenoids present. These entirely mask the small quantity of green chlorophyll.

Dense clusters of small, waxen white flowers precede the production of globose capsules filled with rough or sticky seeds.

Because of the parasitic nature of dodder, chemical treatment is usually lethal to the host plant as well. If cutting out by hand seems too big a job, spraying with 2.5 percent by weight of sulphuric acid or with one pound of ammonium thiocyanate in two gallons of water should kill your dodder infestation—and very likely your desirable host plants as well.

Dodder [*Cuscuta gronovii*]. A. Parasite wrapped around host plant. B. Clusters of small white flowers. C. Single enlarged flower. D. Single enlarged mature ovary. E. Much enlarged seed, which is sticky to help it attach to new host plant. The power of magnification is denoted by X.

D E

X5

B

C

A

X½

Polygonum aviculare L.

COMMON NAMES Knotgrass, matgrass, doorweed, pinkweed, birdgrass, stonegrass, waygrass, goosegrass, allseed, centinode, nine-joints, wine's grass.

SOME FACTS Native to the United States and introduced from Eurasia; reproduces annually by seeds.

BLOOMS June to October.

RANGE Common in northern United States and southern Canada.

HABITAT Hard-trampled lawns, yards, roadsides, paths, waste places.

Whole large sections of many lawns are composed not of grass but of knotgrass, which indicates hard soil, lazy lawn keepers, and careless walkers. At a quick glance, when matgrass has been recently mowed, it gives the appearance of a blue-green grass. Indeed, growing pure stands of knotgrass might be one solution to the suburban lawn grower's problems. (You can bet that some very good grass would immediately invade such a lawn and probably take over!)

Knotgrass leaves are oblong and narrow and can be seen in their bright blue-green color growing happily between the paving stones of the streets of large cities. They are one-quarter to one inch in length. The slender pale green stems are usually prostrate and measure four inches to two feet in length. Branches emerge from the joints (or "knots"), which appear paler within the sheath common to the genus *Polygonum*.

The tiny white to pink blossoms make their appearance either solitarily or growing in groups of two or three within the axils of the leaves. The whole perianth of a single flower is less than two millimeters long!

Though in 1809 William Pitt included knotgrass in his list of the worst weeds of arable land, it has been found to have a number of uses. Its juice stops nosebleed and a tea brewed from it has been used to stop diarrhea and hemorrhages. The American Indians had the patience to collect its tiny seeds and include them with other seeds in a pinole or parched grain. In Shakespeare's day some thought this plant could inhibit the growth of children; thus his "hindering knotgrass," in "A Midsummer Night's Dream." No evidence has been demonstrated for this ability.

If one-half pound per acre of 2,4-D ester or of 2,4-D amine is applied to the seedling stage of knotweed, it will help rid you of this lawn pest. Dicamba *used exactly as the manufacturer directs* should also help. Avoid walking on your lawn and thus packing the soil. Pull knotgrass up before flowering and reseed with a desirable plant.

Knotgrass [*Polygonum aviculare*]. A. A single plant, including root. B. Small leaf. C. Sheath at node—a characteristic of the genus *Polygonum*. D. Single, enlarged flower. E. One fruit. The power of magnification is denoted by X.

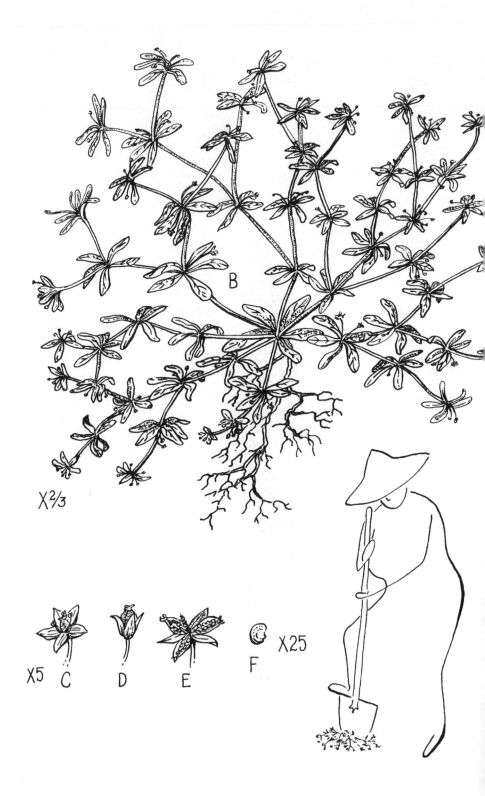

X⅔

X5 C   D   E   F X25
B

Mollugo verticillata L.

COMMON NAMES Carpetweed, Indian chickweed, whorled chickweed, devil's grip.

SOME FACTS From tropical America but native to Africa; annual; reproduces by seeds.

BLOOMS June to September.

RANGE Eastern and middle United States to Florida and Texas.

HABITAT Gardens, lawns, waste places, between the cracks of walks; dry, gravelly, and sandy soil.

While there is not too much that can be said in favor of carpetweed, it can, in an emergency, be eaten, though you'd have to collect a very great deal of this small plant to make a stomach-filling repast!

Its stems are much branched, prostrate, and form mats. Simple, whorled leaves that are entire and spatulate appear in groups of five or six at each node. Small umbel-like flowers are found at the nodes on slender pedicels and in the axils of the leaves (verticillate). These flowers have five sepals, which are white inside and which have taken the place of the petals.

Pull up early, or if you use CDAA (a preemergence herbicide), be careful because it is toxic to grasses.

Carpetweed [*Mollugo verticillata*]. **A.** Prostate plantlet. **B.** Whorled, simple, spatulate leaves. **C.** Single flower. **D.** Single flower with ripening fruit. **E.** Three-valved capsule, open. **F.** Tiny seeds, much enlarged.

C

D

E

F

X4

X4

X4

B

A

X1

Stellaria media Cyrill.

COMMON NAMES   Common chickweed, starwort, starweed, winterweed, birdweed, satinflower, tongue grass.
SOME FACTS   Introduced from Europe; annual; propagates by seeds.
BLOOMS   Throughout the year if winter is mild.
RANGE   Worldwide; very common in North America.
HABITAT   Gardens, lawns, waste places, meadows, cultivated fields.

Don't be fooled by the small size and delicate structure of chickweed. It is one of the commonest weeds in America, if not the commonest. Wherever the stems touch the ground (and they generally lean on the ground), the nodes give rise to roots and new stems. Furthermore, the plant continues to make new seeds right through a mild winter, or gets a very early start in the spring if the winter was a difficult one. That combination of features makes the winterweed a most successful plant, and in the two hundred years or so since it arrived fom Europe, it has spread across the land.

Chickweed is a member of the pink family and is related to carnation and bouncing bet (p. 213). The fundamentally five-parted flowers appear in the leaf axils or in cymose clusters, their white cleft petals being shorter than their green sepals. A small one-chambered capsule is produced in which tiny, approximately circular seeds are found, each measuring no more than one millimeter in diameter.

The low, slender branches of starweed are much branched and covered with rows of hairs. Opposite leaves are produced, and they are ovate to oblong in shape. The higher leaves are largely sessile, while those lower down on the stem have hairy petioles.

This ubiquitous weed makes a fine potherb. In more ancient days it was used as a poultice for boils, inflammation, and external abscesses. One old wives' tale has it that eating this plant will reduce body fat. If this were widely believed, the weed would probably vanish from the face of America! However, I don't think the claim is true. Chickens feed on it, as its most common common name suggests, and they fatten.

Chickweed is moderately sensitive to MCPA and 2,4-D, and quite sensitive to mecoprop salt or dicamba. Sodium nitrate or sodium chloride will also kill chickweed. It is difficult to eradicate by hand because it has rooted at every spot where a node touched the soil and it is easy to miss some of the rootings.

Common chickweed [*Stellaria media*]. **A.** Whole plant showing the root. Note the roots that spring up at the nodes where stems touch the soil. **B.** Approximately ovate leaves. **C.** Small white flower. **D.** Single flower enlarged, showing the pinked petals. **E.** Single flower unopened, showing both sepals and petals. **F.** Single seed enlarged. The power of magnification is denoted by X.

Fragaria virginica Duchesne.

COMMON NAMES Wild strawberry, strawberry.
SOME FACTS Native; perennial; reproduces by seeds and runners.
BLOOMS April to July.
RANGE Common in northeastern United States.
HABITAT From Labrador and Newfoundland to Georgia.

Our own early American botanist William Bartram, traveling along the little Tennessee River in the early 1770s (which means he was still a British botanist), had this to say of Alabama in general and strawberries in particular: "Having gained its summit, we enjoyed a most enchanting view—a vast expanse of green meadows and strawberry fields, a meandering river gliding through, saluting in its various turnings the swelling, green turfy knolls embellished with parterres of flowers and fruitful strawberry beds, flocks of turkeys strolling about them, herds of deer prancing in the meads or bounding over the hills, companies of young, innocent Cherokee virgins, some busy gathering the rich, fragrant fruit. Others, having already filled their baskets, lay reclined under the shade of floriferous and fragrant native bowers of magnolia, azalea, perfumed calycanthus, sweet yellow jessamine, and cerulean glycine disclosing their beauties to the fluttering breeze and bathing their limbs in the cool, fleeting streams; whilst other parties, more gay and libertine, were yet collecting strawberries, or wantonly chasing their companions, tantalizing them, staining their lips and cheeks with the rich fruit." It makes a delightful picture, doesn't it?

Interestingly, the strawberry is not a berry at all. Botanically, a berry is a *fruit*, i.e., a ripened ovary. What is consumed with such pleasure when one eats strawberries is the receptacle on which sit the pistils. The receptacle is a modified portion of the stem that, in the case of the strawberry, turns sugary and moist when mature. The true fruits are the tiny dots that are embedded in the surface of the "berry." (If they are embedded, you are eating the fruit of *Fragaria virginica*, but if they are flat against the surface of the receptacle, you are consuming the fruit of *F. vesca*.) The real fruit of strawberry is called an achene.

The little white flowers are very similar to the yellow ones of *Potentilla*. Both genera are very closely related members of the rose order (Rosales). Clusters of as many as twelve flowers may be found on pedicels of approximately equal length. Trifoliate toothed leaves are borne on longish petioles.

Dig out the plantlets and be sure to get the runners.

Wild strawberry [*Fragaria virginica*]. A. Whole small plant of strawberry. Note the roots that are springing up where runner touches the ground. B. Leaf composed of three toothed leaflets. C. Small five-parted flowers. D. "Fruits" beginning to grow. E. Single enlarged flower with five green sepals, five white petals. Observe the many separate pistils on the long receptacle. Here the receptacle becomes the fruit of commerce by becoming sweet and soft. F. A "fruit" maturing. G. The true fruit, an achene (barely visible on F).

Trifolium repens L.

COMMON NAMES White clover, ditch white clover, white trefoil, purple grass, purplewort, honeystalk, lamb suckling, honeysuckle clover, trinity leaf, shamrock, husbandman's barometer.

SOME FACTS Native of Europe (and possibly present in northern North America before the colonists came over); perennial; reproduces by seeds and has a creeping stem.

BLOOMS All summer long.

RANGE Newfoundland to Alaska and south to Florida and California, but more abundant in north.

HABITAT Fields, lawns, copses.

Is there anyone with soul so dead as never to have hunted for a four-leaf clover? Next time you find one, keep in mind that you are seeing before you four *leaflets* of one leaf and not four leaves. Why is it lucky to find one? This goes back to the time when any symbol for the cross was luck-inducing, and the four leaflets form a stubby cross. Indeed, the plant is also considered holy when it has but three leaflets. It was St. Patrick himself (it is rumored) who used the three-parted leaf to demonstrate the doctrine of the Trinity to the Irish. Clover leaves are reputed to be quite noisome to witches (no mention of warlocks).

Each leaflet of the leaf is notched at the outer end (obcordate), and you will find that the leaflets fold at night or at the approach of a storm when the sky darkens. This helps account for the common name of husbandman's barometer.

The small flowers of white clover are pollinated by bumblebees, which take nectar (which is over 40 percent sugar) from the small, numerous flowers of the round head. After pollination, each flower bends downward and then turns brownish while the fruit and seed mature.

The stems root at every node, which accounts for the very rapid spread of white clover (it will very soon fill all the spaces of your bluegrass lawn).

It is interesting that the leaves have been found to be cyanogenetic yet the species has been used for generations as cattle feed. Mathias L'Obel, a sixteenth-century Flemish herbalist, reported that the plant was good for fattening cattle and suggested that it be used to fatten peasants, too. The lovely lobelia is named for L'Obel.

Though *Trifolium repens* is resistant to MCPA and to 2,4-D amine, it is quite sensitive to mecoprop.

Purple clover [*Trifolium pratense*] and white clover [*T. repens*]. A. Purple clover flowering stalk and leaves, with three leaflets, a white V on each leaflet. B. Head of papilionaceous flowers. C. Individual flower. D. Nodules on root containing nitrogen-fixing bacteria. E. Leaves and flower heads of white clover. F. Turned-down pollinated flowers. The power of magnification is denoted by X.

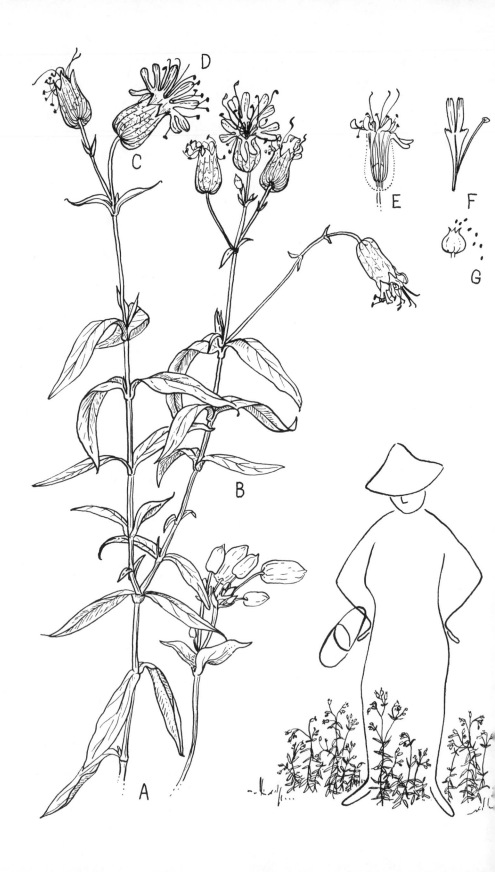

Silene cucubalus Wibel

COMMON NAMES Bladder campion, bubble poppy, white bottle, cowbell, snappery, devil's rattle box, maiden's tear.

SOME FACTS Perennial; reproduces by seeds and rootstocks; originally from north Africa; came to United States through Europe and was naturalized here.

BLOOMS June to September.

RANGE Common now in the East and widely distributed throughout temperate North America; local in Pacific Northwest.

HABITAT Waste places, grasslands, roadsides, cultivated ground, meadows.

The most striking thing about bladder campion is its "bladders." These are composed of the thin-textured sepals, fused to form the urn-shaped bladder. This structure has about twenty pinkish interwoven veins, which are quite observable. The small urn-shaped capsule is provided with six teeth at its top and opens there as well.

Leaves that are opposite and sessile are found along the stem and each tapers to its tip. Each of the five petal edges is divided into two lobes, suggesting ten rather than five petals.

The young plants are palatable when cooked—they taste like peas, though they are somewhat bitter because of the presence of saponin, a chemical common in members of the pink family. Saponins lather in water, but in large amounts are dangerous because they dissolve the red blood cells. *Saponaria officinalis* (p. 213), our soapwort, has leaves with much saponin in them. Cooking destroys much of the saponin. In 1685, when most of the crops on the island of Minorca failed, the population fell to dining on bladder campion. It also makes a perfectly good fodder.

It is resistant to many herbicides and is only moderately sensitive to mecoprop salt. Hand pulling is probably the easiest way to remove bladder campion should you want to.

Bladder campion [*Silene cucubalus*]. **A.** Stalk with flowers. **B.** Opposite, sessile leaves. **C.** The swollen "bladder," which is formed from fused sepals. Venation can be made out. **D.** The lobing of the petals makes it appear there are ten petals when there are but five. **E.** A logitudinal view showing (dotted line) where the bladder is. **F.** A single bilobed petal showing that the stamens are fused to the petals. **G.** The small urn-shaped capsule shedding tiny seeds.

F

G
X10

D

C

B
X⅓

A ♀
X¾

Lychnis alba Mill.

COMMON NAMES White campion, white cockle, evening lychnis, snake cuckoo, thunder flower, bull rattle, white robin, mother-die.

SOME FACTS Introduced from Europe; perennial or biennial; reproduces by short rootstocks and seeds.

BLOOMS June to August.

RANGE Eastern United States and Canada; locally common in north central states and Pacific Northwest.

HABITAT Waste places, lawns, grain fields.

If your intentions are matricidal, pick a flower from the mother-die plant. It was once thought that picking a blossom from white campion would result in the death of the picker's mother. (Picking the red-flowered campion, *Lychnis dioica*, was once thought to be lethal to the picker's father.) As if this is not bad enough, there is a distinct danger that should you pick one of the white flowers you will soon be hit by lightning.

White cockle's spreading, branching stem rises to two or three feet and is hairy and slightly sticky, particularly in its uppermost parts. Its simple, opposite leaves are ovate-lanceolate and also hairy. The lower leaves taper to margined petioles, while those higher on the stem are sessile.

The five-petalled flower may be staminate or pistillate; staminate flowers occur on separate plants from those bearing pistillate flowers. Both types of flowers occur in loose cymose clusters, though solitary flowers are frequently seen in the field.

The five-fused sepals of the pistillate flower form an inflated calyx that is red-tinged along the hairy veins. The calyx is less swollen in the staminate blossom. All five white or pink petals of the fragrant inch-wide flowers are notched on their outer edges, which may falsely suggest more petals than are actually present. Five styles may be seen at the top of the floral tube. (This may help you separate this flower from its near relative, *Silene noctiflora*, which has three styles.)

The capsule contains numerous seeds and splits open by ten teeth at its top.

Do not allow to seed. Remove by pulling, being sure to get the thick, fleshy taproot.

White campion [*Lychnis alba*]. A. Flowering shoot with opposite leaves and flowers at the top. B. The taproot. C. Leaves with three major veins visible. D. Five-parted flower with each petal divided into two lobes. E. Swollen base of flowers. F. Mature fruit shedding the tiny seeds. G. A much enlarged seed. The power of magnification is denoted by X.

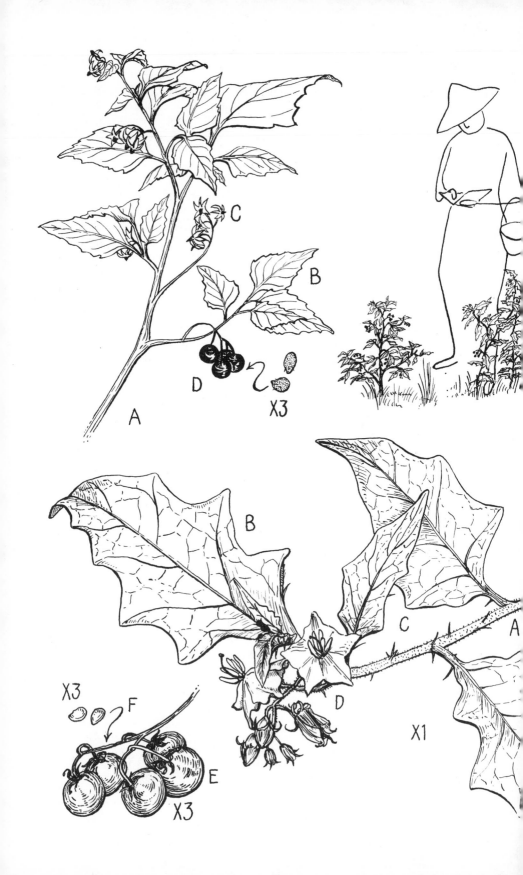

Solanum nigrum L.

COMMON NAMES Black nightshade, deadly nightshade, poisonberry, hound's berry, stubble berry, garden nightshade.

SOME FACTS Native to the United States; annual; reproduces by seeds.

BLOOMS July to September.

RANGE Common all over the United States.

HABITAT Fields and waste places; common on loamy or gravelly soils.

Deadly nightshade is a frightening common name indeed, and reports do indicate that the fruits are poisonous, since they contain solanine, a toxic glyco-alkaloid. Many members of the potato family—including potato, tomato, eggplant, ground cherries, and jimson weed (p. 197)—produce chemicals of a poisonous nature, e.g., nicotine or atropine.

The roughly rhombic, toothed leaves are alternately displayed upon a stem one to two feet high, which is often much branched. The leaves are not armed as they are in the closely related horse nettle (also called wild tomato), *Solanum carolinense*. In the latter species the leaves sport stout yellowish prickles along their veins, midrib, and petiole.

Deadly nightshade's small, white, relatively wheel-shaped flowers have five petals, which are produced in small, umbel-like drooping clusters at the side of the stem but near its top. (These blossoms are very similar to those of horse nettle, *S. carolinense*, but smaller.) The berries that form following fertilization turn from green to black. They are occasionally baked into pies, and no deaths or illnesses seem to follow eating of nightshade-berry pie. Indeed, some Midwestern housewives swear by the berries, and in some places the plant is called "wonderberry." Taxonomists now recognize differences between *S. nigrum* and the wonderberry; they have renamed the latter *S. intrusum*. Whether or not its berries are poisonous, deadly nightshade's leaves and stems definitely are not and may be safely used as a potherb with proper cooking.

Nightshade is sensitive to simazine but resists 2,4-D. If you decide to pull up any horse nettle on your property, be sure to wear thick gloves and beware the armament.

(TOP) Deadly nightshade [*Solanum nigrum*]. A. Portion of shoot. B. Leaf showing dentition and pinnate venation. C. Small white flowers with petals bent back slightly. D. Blue-black berries and a few enlarged seeds. (BOTTOM) Horse nettle [*Solanum carolinense*]. A. Portion of a shoot. B. Larger armed leaf. C. Stem armament (prickles). D. Single five-parted flower. E. Enlarged berries. F. Enlarged seeds.

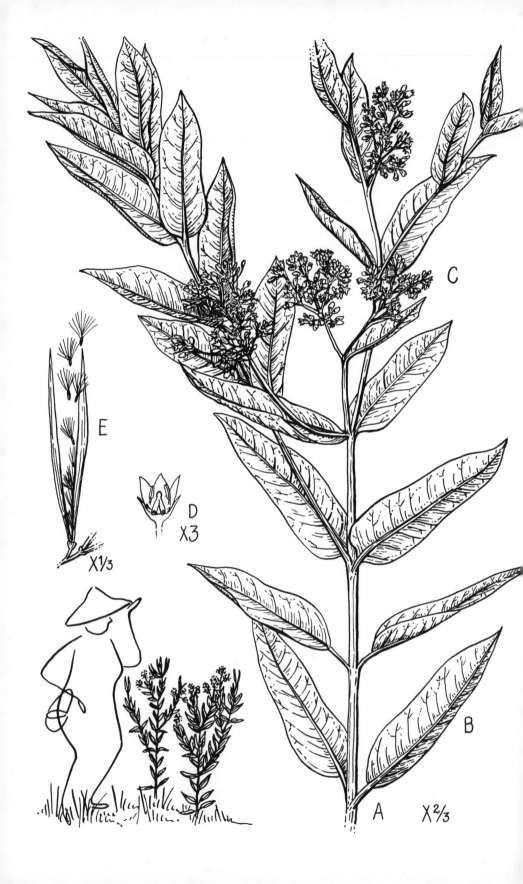

A X⅔

B

C

D
X3

E

X⅓

Apocynum cannabinum L.

COMMON NAMES Dogbane, American hemp, Indian hemp, Choctaw root, bowman root, dropsy root, blind hemp, rheumatism weed.

SOME FACTS Native; reproduces by seeds and rootstocks; perennial.

BLOOMS June to August.

RANGE Widespread in the United States.

HABITAT Fields, thickets, moist soils.

Despite the common name dogbane, dogs do not go out of their way to avoid this plant; indeed, the name applies to its European relative rather than to our species. Most animals avoid eating the plant because it is distasteful.

Dogbane's one- to five-foot-long branching stem, whose main axis is usually exceeded by the branches, arises from a deep, vertical, and branched root. From the root are extracted several resins and glycosides, some of which act as heart stimulants and diuretics.

Oblong, opposite, two- to four-inch-long leaves, seemingly pointed at both ends, emerge from the stems on short petioles or are sessile. If a leaf is removed, a milky juice will appear at the break.

The terminal or axial flowers, in dense clusters that are flat-topped, are not strikingly colorful or large. Each blossom is five-lobed and greenish-white. The twin four-inch pods (follicles) that result from fertilization are round and smooth, and while maturing, relatively conjoined at their distal ends, though they can be easily separated by a touch. Each pod is filled with numerous brownish flat seeds attached to a tuft of fine white hairs. Thus parachuted, the seeds are carried far and wide by the wind.

This is what Kalm reports of Indian hemp: "*Apocynum cannabinum* was by the Swedes called Indian hemp, and grew plentifully in old grain grounds, in woods, on hills, and in high glades. The Swedes have given it the name Indian hemp, because the Indians formerly and even now apply it to the purposes as the Europeans do hemp: for the stalk may be divided into filaments and is easily prepared. When the Indians were still living among the Swedes in Pennsylvania and New Jersey, they made ropes of this *Apocynum*, which the Swedes bought, and used them as bridles, and for nets. The Swedes usually got thirty feet of these ropes for one piece of bread[!]."

From the much-branched root is obtained an extract containing several resins and glycosides, some of which have a stimulating effect on the heart (hence dropsy root).

Don't let this plant remain on your land too long. Its spreading rootstocks, as well as its abundant wind-carried seeds, mark it as a weed that must be rooted out at once. Spraying small patches of it with sodium chlorate may help.

Dogbane [*Apocynum cannabinum*]. **A.** Portion of shoot. **B.** Oblong leaf, which, if broken from the stem, will ooze milky juice (latex). **C.** Clusters of small flowers. **D.** Single flower enlarged. **E.** A pair of joined follicles.

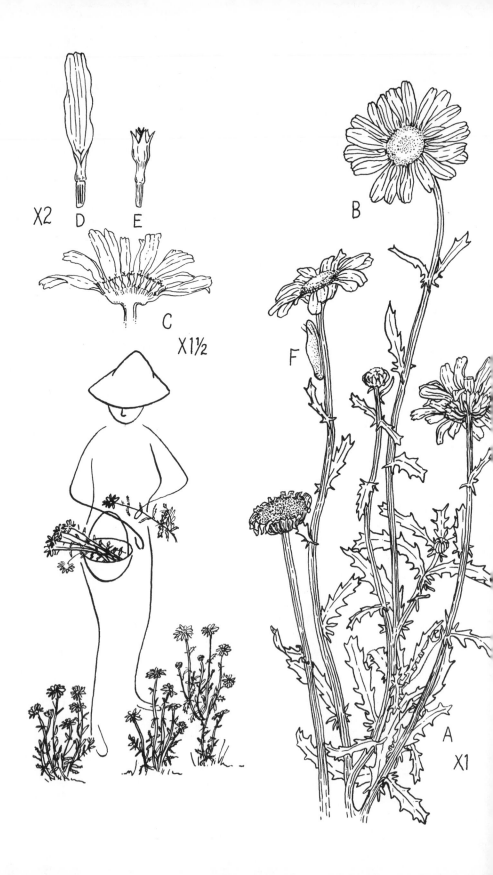

X2   D   E

B

C

X1½

F

A

X1

Chrysanthemum leucanthemum L.

COMMON NAMES Ox-eye daisy, white daisy, white weed, field daisy, poor-land flower, marguerite, poverty weed, moon-penny, dog blow, maudlin daisy, bull daisy.

SOME FACTS Introduced from Europe; perennial; reproduces from seeds, and not very effectively from short rootstock.

BLOOMS June to July.

RANGE Common in northeastern United States; also found southward, and westward to the Pacific Coast.

HABITAT Waste ground, oil fields, pastures, gardens.

Ox-eye daisy is a very well known and liked plant; indeed, it has been proclaimed the state flower of North Carolina. Who has not gathered its white and gold "blossoms"? Who has not plucked the "petals" one by one, asking those fateful questions? Yet, this same plant, if given a chance, becomes a very noxious weed. Poorland flower will take over a pasture with alacrity, much to the owner's chagrin. Its spreading rootstocks, as well as the many seeds it makes, aid its travel.

The twenty to thirty white "petals" are, since the plant is a member of the Compositae, not individual petals at all. Each petal is composed of five of these false petals fused together. If you study their edges closely, you will be able to see a waviness, which indicates the outline of the five. In addition to the white ray flower, there is the golden center, or "eye," composed of tubular disc flowers, each, in turn, composed of five hairlike sepals, five fused petals, stamens, and two fused pistils, a complete flower.

The head (or capitulum, which means little head) is terminal on a one- to three-foot-long stem. Several stems may be produced from each root crown, and each has simple, alternate leaves whose edges are toothed. The leaves along the stem differ from those of the rosette at the base of the stem in that the higher leaves are sessile, while those of the rosette are petiolate and pinnately-lobed as well as oblong or lanceolate.

Marguerite's young leaves can be used in salads, and the plant has also been used for home remedies for whooping cough, asthma, and other coughs. A tea made from its leaves is sometimes used as an antispasmodic. Gerard says of it, "The juice and the leaves and roots shift up into the nostrils, purgeth the head mightily, and helpeth the megrim. The same given to little dogs with milke, keepeth them from growing great."

Poorland flower is moderately resistant to MCPA and to 2,4-D, and even to dicamba. The best thing to do is to dig out the plant before the heads are produced. Be sure to get the underground parts.

Ox-eye daisy [*Chrysanthemum leucanthemum*]. **A.** Flowering stalks showing alternate, small, narrow pinnately lobed leaves. **B.** Single silver-dollar-sized head. **C.** Single head cut in longitudinal section to reveal the ray flowers at the side and the disc flowers at the center. Disc flowers are yellow and ray flowers are white. **D.** Single ray flower. **E.** Single disc flower. **F.** Garden slug reputed by some to pollinate this plant.

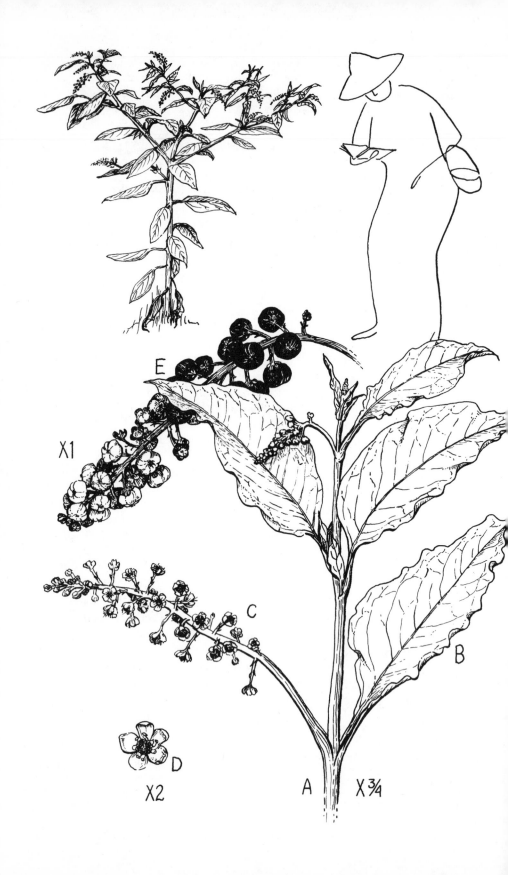

E

X1

C

X2   D   A   X¾   B

Phytolacca decandra L.

COMMON NAMES Pokeberry, inkberry, poke, red-ink plant, scoke, pigeon-berry, garget, pocan bush, charges.

SOME FACTS Native; European gardeners have adopted it; perennial; reproduces by seeds and rootstocks.

BLOOMS June to September.

RANGE Common in nearly all of temperate North America.

HABITAT Waste places, fence rows, thickets.

Many people in the United States still consider this large weedy plant a valuable food plant. Its young leaves taken from the eight- to ten-inch shoots make a fine spinach or greens substitute, but the water used in cooking them should be changed at least twice and thrown away. Young shoots taken before the appearance of woodiness taste, after proper cooking, much like asparagus. The American Indians were very fond of the plant as a food.

Inkberry produces a root of large size that contains phytolaccin, a cathartic. Pigs have been poisoned from ingesting pieces of the root of pokeberry; they died of bleeding gastritis. Man can also be poisoned by pokeberry root.

The purple-red berries contain a fugitive dye. The berries are occasionally made into pies, with no recorded disastrous results for the diners, though the seeds are supposed to contain poisonous principles.

Poke may grow to be ten feet tall. The older stems turn purple later in the season. These stout stems may branch above, and are covered with alternate, simple, entire leaves with long petioles. Fundamentally of an oblong-lance shape, some may reach the length of one foot, and when bruised, emit a somewhat unpleasant odor.

Smallish white flowers are borne in racemes; each flower has ten stamens, the source of the specific epithet (*deca*, ten; *andra*, male organ).

The American Indians used pokeroot decoctions (as did the early American settlers) for a variety of illnesses. Kalm once more wrote about this weed: "Mr Bartram mentioned having hit his foot against a stone, he had gotten a violent pain in it; that he had then bethought himself of putting a leaf of the phytolacca on his foot, by which he had lost the pain in a short time and got his foot well soon after."

Cut off the main stem above ground level and then apply dry salt, carbolic acid, and kerosene to the cut surface. The rhizome must be destroyed or *P. decandra* will return year after year.

Pokeberry [*Phytolacca decandra*]. **A.** Portion of growing shoot. **B.** Single oblong lance-shaped leaves. **C.** Single raceme of small flowers. **D.** Single flower enlarged, showing five petals and ten stamens. **E.** The raceme of maturing berries. Older berries lower down. The power of magnification is denoted by X.

A

B

C

D

E X10

X ¾

Datura stramonium L.

COMMON NAMES Jimson weed, Jamestown weed, thorn apple, mad apple, stinkwort, angel's trumpet, devil's trumpet, stinkweed, dewtry, white man's weed.

SOME FACTS The origin of this weed is in doubt—some think it originated in India, and others that it was introduced from the tropics; annual; reproduces by seeds.

BLOOMS June to September.

RANGE Florida to Texas, and north to Canada.

HABITAT Fields, waste places, dumps.

The colonists of Jamestown, Virginia, soon came to know this large weed; it was given the name Jamestown weed, later corrupted to jimson weed. In 1676, soldiers sent to quell Bacon's Rebellion in Jamestown were poisoned by it, much to the delight of Bacon, no doubt. In Dioscorides' *Greek Herbal*, written in the first century A.D., he wrote of the thorn apple: "The root being drank with wine ye quantity of a dragm, hath ye power to effect not unpleasant fantasies. But 2 dragms being drank, make one beside himself for three days & 4 being drank kill him." Enough said.

Jimson weed is a large and strikingly flowered plant that rises to a height of five feet when growing under happy conditions. Its pale green stems branch by forking; attached to them are alternate, large, oval but irregularly lobed and toothed ill-scented leaves. Attached by stout petioles, they are darker green above than below, and are large-veined.

Within the forking of the branches appear the solitary three- to four-inch-long trumpet-shaped attractive flowers on a short peduncle. Each of the five white petals has a prominent tooth. A five-lobed and ridged calyx surrounds the petals, reaching halfway up the floral tube. Flowering is followed by the production of two-inch-wide oval, spiny capsules, which when dry open at the top. These contain four cells and many dark brown, flat, kidney-shaped, three-millimeter-long, and very poisonous seeds.

Every part of this plant is potentially lethal. Several powerful alkaloids are found in its tissues and all are poisonous when consumed in fairly small quantities. (Four to five grams of crude leaf extract is enough to kill a child.) Still, they are very useful medicinally when taken in trace quantities.

The tissues of this plant contain the alkaloids scopolamine and atropine among others. Scopolamine is used by doctors as a preanesthetic in childbirth, surgery, and ophthalmology. Even the hand that crushed a leaf when touched to the eyes should cause dilation of the pupils.

Before it seeds, pull it out and wear gloves. The seedlings are sensitive to MCPA salt and 2,4-D amine.

Jimson weed [*Datura stramonium*]. **A.** Portion of flowering shoot. **B.** Large pinnately lobed leaves. **C.** Large trumpet-shaped flower showing the five fused sepals and petals. **D.** Fruit, a capsule, is large and heavily armed (will cut fingers). **E.** A much enlarged seed from the capsule.

D

E

X½

X4

F

X4

C

A
X1

B

Eupatorium perfoliatum L.

COMMON NAMES   Boneset, thoroughwort, ague-weed, fever weed, sweating plant, crosswort.

SOME FACTS   Native to our shores; perennial; reproduces by seeds.

BLOOMS   August to September.

RANGE   Widespread in eastern North America, westward to Nebraska and Texas.

HABITAT   Wet fields, pastures, swamps, sides of streams and ditches.

According to the Doctrine of Signatures, one had only to find a plant with a leaf shaped like a heart in order to find a cure for heart ailments. It was assumed that any plant organ that looked like a human part was capable of yielding a remedy for the sickened human part. Since the opposite, wrinkled, pointed leaves of boneset are united at their bases, the stem passing through the area of fusion (*per*, through; *folia*, leaf), it was obvious to our ancestors that a poultice made of the leaves and applied to a broken bone would help the parts rejoin. Oh, for the restoration of such optimism! However, boneset is still used as a home remedy to dissipate a cold by inducing sweating.

The fused leaves are found on a two- to six-foot-high rather stout stem that branches at the top. Unlike its close relative, Joe-pye weed (p. 231), the compact corymbose clusters of small perfect flowers are a dull white, though occasionally one does find a boneset with blue flowers.

Hand pull if the plant is unwanted.

Boneset [*Eupatorium perfoliatum*]. A. Portion of the stout shoot. B. Large leaves, which taper to a point, are dentate and rough to the touch. They seem to be fused about the stem (perfoliate). C. Small flowering heads clustered at top of plant. D. Single head. E. Single ray flower. F. Single seed with parachute. The power of magnification is denoted by X.

X1

F
X2

X4
G

E
X4

X1

D

C

A

B

X1

Lepidium virginicum L.

COMMON NAMES Peppergrass, bird's-pepper, poor man's pepper, tongue grass.
SOME FACTS Native; annual; propagates by seeds.
BLOOMS May to September.
RANGE Newfoundland to South Dakota, southern Florida to Texas.
HABITAT Waste places, roadsides, dry soils.

Lepidium sativum, a European relative of our native peppergrass, is a salad plant well known for its peppery but interesting taste. Though ours is peppery too, it is unfortunately not as tasty, except possibly to birds, which devour the small fruits by the thousands (thus giving the plant one of its common names, bird's-pepper).

The much-branched stem is six to twenty-one inches in height. The lower leaves are pinnatifid, approximately spoon-shaped in outline, their terminal lobes large and their lateral lobes very small. Leaves that are higher on the stem are lance-shaped and smooth.

The elongating racemes of white-greenish four-petalled flowers produce, after fertilization, two-celled pods (siliques), which are smaller and rounder than those of the plant's cousin, *Capsella*. These tiny pods are also flat and notched at their upper end, but unlike their many-seeded relative, they contain only two reddish-yellow seeds.

Lepidium often breaks off at the base of the stem and rolls with the wind, scattering its seeds far and wide—acting, that is, like a tumbleweed.

Getting rid of peppergrass is accomplished the same way as getting rid of *Capsella* (p. 203).

Peppergrass [*Lepidium virginicum*]. A. Portion of stem showing. B. Deeply toothed flowering stalk leaves. C. Racemes of maturing fruits. D. Tiny flowers. E. Single flower showing four petals (cruciform in shape). F. Single inflated, almost heart-shaped fruit (silique) containing two seeds. G. One of the seeds much enlarged. The power of magnification is denoted by X.

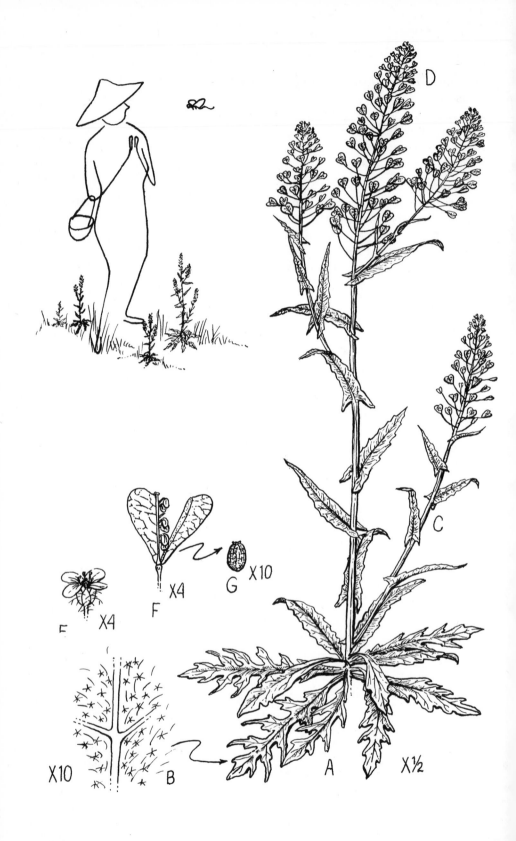

X4

X4

F

X10

B

X10

G

F

X4

C

D

A   X½

Capsella bursa-pastoris (L) Medic.

COMMON NAMES Shepherd's purse, pepperplant, caseweed, pickpurse.
SOME FACTS Introduced from Europe; annual; reproduces by seed production.
BLOOMS March to November in northern United States; all year elsewhere.
RANGE Common throughout the United States.
HABITAT Gardens, cultivated fields, waste places.

Shepherd's purse is apparently bitter to an animal's taste, but it has never been found to account for the poisoning of either animals or man, which is fortunate since it is one of the most common of the weedy plants in the United States. Three hundred years ago it was granted to have some medicinal value. The herbalist Gerard said of it, "Shepheard's purse staieth bleeding in any part of the body whether the juice or the decoction thereof be drunke, or whether it be used pultesse wise, or in bath, or any other way else."

Capsella bursa-pastoris is one of the many members of the Cruciferae family, and therefore related to mustard, the wallflower, stock, honey, candy-tuft, peppergrass (p. 201), cabbage, radish, and other valuable edible plants, and to many other garden plants. It has been known to harbor the fungal disease that produces club root in cabbage, cauliflower, and radish, and is therefore unwanted near farmland.

A rosette of alternate, simple, toothed leaves appears first on the surface of the ground, while under the soil is a thin taproot. From the rosette's center arises a thin, erect, branched stem with leaves, but these leaves are unlike those of the rosette. They clasp the stem and are arrow-shaped because of the small pointed ears at their bases. At the tip of the branch a slender lengthening raceme of small, four-petalled white flowers, with the typical cruciform petal arrangement, is produced.

Flowers continue to open at the top of the raceme, while mature (triangular) fruits (siliques) are found lower down on the same stem.

The fruits look like little boxes (*capsella* means little box) or like the little purses (*bursa* means purse) pastoral maidens once carried about with them. At maturity the fruits open to shed several to many seeds, each seed being only about one millimeter in diameter. A single plant may produce two thousand or more seeds each year.

Shepherd's purse is sensitive to a number of weed-killing chemicals such as MCPA salt, 2,4-D amine, and simazine. Careful spraying with iron sulphate or copper sulphate should also do these small plants in.

Shepherd's purse [*Capsella bursa-pastoris*]. A. Habit of weed showing almost pinnatifid basal leaves. B. Very enlarged drawings of the stellate hairs on the lower leaves. C. The arrow-shaped leaves of the flowering stalk, which are sessile. D. A raceme of flowers (at the top) and maturing fruits (lower down). E. Single four-parted flower (Maltese cross formed by petals). F. Single mature fruit opening. G. Much enlarged seed.

A   X²⁄₃   D   E   C

F X2   G X2   H X2   X½   B

Polygonum cuspidatum Sieb and Zucc.

COMMON NAMES Japanese knotweed, Mexican bamboo.
SOME FACTS Native of Japan; perennial; propagates by seeds and by rhizomes.
BLOOMS August to September.
RANGE Widely found in the northeastern part of the United States, west to Minnesota and Iowa, south to Maryland.
HABITAT Waste places, neglected gardens.

In 1864 a Belgian botanist brought Japanese knotweed from Japan to England and shortly thereafter this rapidly growing plant arrived on American shores. It must be admitted that a large stand of Mexican bamboo is attractive in a masculine sort of way, but it is also a potential danger; *Polygonum cuspidatum* can take over a garden in very short order.

Mexican bamboo is not a member of the grass family (true bamboo belongs to the grasses), but is a dicotyledonous plant and a member of the buckwheat family, to which belong *Rumex acetosella* (p. 215), buckwheat, and rhubarb. However, the three- to nine-foot-tall stems do resemble bamboo stems superficially and in their rapid growth. Stems of Mexican bamboo are also hollow at their centers.

The underground stem or rhizome, which may attain a length of five or more feet, makes this persistent plant difficult to root out. In the fall the erect portion dies back to the rootstock, and this adds to the ugliness of *P. cuspidatum* in a garden.

The plant's leaves are broadly ovate, relatively truncated at their bases, and attached to the stem by petioles. Around the nodes of all members of the genus *Polygonum* are sheaths which may be described as papery, cufflike masses of tissue.

The apetalous flowers are small and greenish-white and are found in branched panicles. Though small, the sepals are showy floral organs and, interestingly for a dicot, there are three fused pistils with three separate stigmas. As the fruit forms, the three lower membranous sepals enclose it.

There are a few good things about Japanese knotweed, all of them culinary. The young shoots are tasty when boiled, resembling asparagus.

The plant is resistant to simazine, boric oxide, and sodium chlorate. Dicamba or picloram may be used with some success (about one-third of an ounce per thousand square feet of land). Use only in the late fall or before May. Many weed killers will kill the foliage but leave the rhizome in fine fettle, in which case you have only scotched the snake. A double layer of black polyethylene laid over a large patch and held down with stones or flowerpots should eventually starve the rhizome.

Mexican bamboo [*Polygonum cuspidatum*]. A. Portion of shoot showing its zigzag bending. B. Portion of the rhizome that makes this plant so hard to kill. C. Large, somewhat cordate leaves. D. Small drooping panicles of white flowers. E. Nodal sheath common in the genus *Polygonum*. F. Single five-parted flower. Sepals have the function of petals. G. Flower in side view. H. Single fruit.

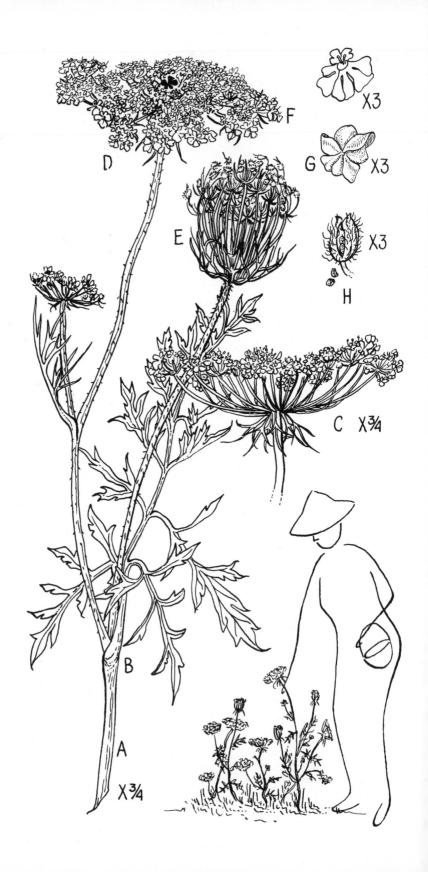

D

F

E

C X¾

A

B

X¾

G X3

H X3

X3

Daucus carota L.

COMMON NAMES Wild carrot, Queen Anne's lace, bird's-nest weed.
SOME FACTS Introduced from Eurasia; biennial; propagates by seeds.
BLOOMS June to September.
RANGE Common throughout North America.
HABITAT Waste places, meadows, pastures, roadsides.

During the reign of Good Queen Anne, lace collars were delicate and lovely, and the lacy, delicate compound inflorescence of wild carrot resembles the patterns of those collars; hence the common name Queen Anne's lace. Held close to the eye, each individual, tiny umbellet of the compound umbel is seen to be composed of a number of small white flowers, and at the center of the entire inflorescence blooms a single flower, almost black in its deep purpleness.

After pollination, during seed set, the outer umbellets bend inward, causing the whole inflorescence to appear concave and look much like a bird's nest. It is during seed maturation that an oil, smelling something like turpentine, can be extracted from the inflorescence.

Though called wild carrot and possessed of a sturdy long white taproot, even under cultivation the plant will not produce the sweet, thick orange taproot known to commerce as the carrot. The commercial carrot is of the same genus and species as wild carrot, but it is of a different strain. It escaped early on from cultivation.

A crown of lacy leaves is also produced during the first year's growth, each twice or three times pinnate, giving the rosette a feathery appearance. The following year the flowering stalk makes its appearance and bears relatively few sessile or clasping leaves. If crushed, the leaf gives off a characteristic and not too unpleasant odor. Boiling water poured over dried carrot leaves produces a carrot leaf tea that some herbalists say has diuretic properties. The seeds are also supposed to be good in preventing flatulence. It has even been suggested that carrot leaf tea will improve the disposition.

No evidence of poisoning has been recorded in America. Animals generally avoid eating the plant, but this is probably because of the aromatic properties of its leaves. Wild carrot *is* closely related to one of the most poisonous plants on earth, the poison hemlock (*Conium maculatum*) given by the Greek state to Socrates. (The hemlock is not the same as the lovely conifer).

Wild carrot is resistant to moderately resistant to most standard weed killers. MCPA or 2,4-D amine may have to be applied repeatedly. Dig out the first year's rosettes. If you miss them, be sure to cut off the flowering stalk before the flowers set seed.

Wild carrot [*Daucus carota*]. A. Second year's growth of this biennial, the flowering stalk. B. Once pinnately dissected leaves. C. Single large umbel in side view. D. Single large umbel showing one dark purple flower at center. E. Umbel after flowers have been fertilized. F. Single white flower of umbel. G. The one purple (almost black to the eye) central flower. H. Fruit opening with small seeds emerging.

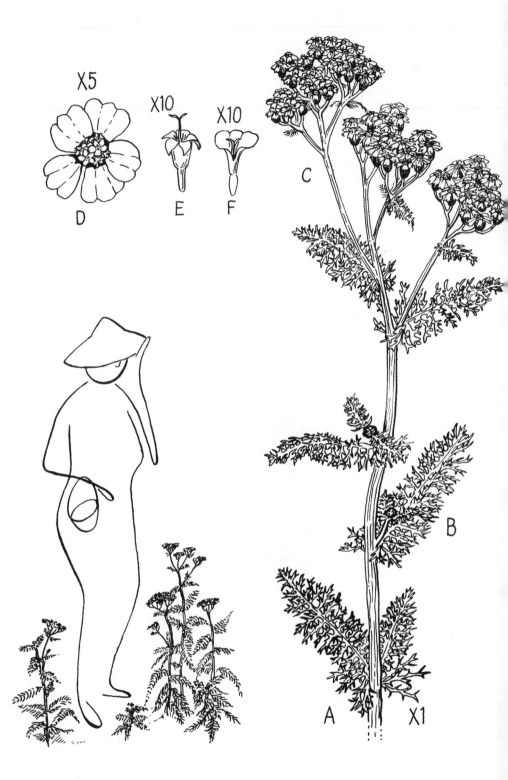

X5

X10

X10

D

E

F

C

B

A

X1

Achillea millefolium L.

COMMON NAMES Yarrow, milfoil, knight's milfoil, thousandleaf, bloodwort, soldier's woundwort, nosebleed weed, devil's nettle.

SOME FACTS Native; perennial; propagates by seeds and by rootstocks.

BLOOMS June to October.

RANGE Throughout most of North America and in most parts of the world.

HABITAT Waste places, meadows, pastures, roadsides.

Achilles is said to have used soldier's woundwort to staunch his soldiers' wounds at Troy, and indeed he may have, for yarrow has definite astringent properties, making the common name woundwort a truly descriptive one. In addition, it is said that if some yarrow is sewn into a flannel cloth and placed beneath one's pillow, one will dream of one's future beloved.

Thousandleaf stands one to three feet tall and is sometimes branched above. Though each leaf appears to be divided into a thousand parts, the number of tiny filaments composing each leaf is far smaller than that. Each sessile leaf is bipinnately dissected into fine divisions, and thus is easily mistaken for a leaf divided into a thousand parts. It also causes the plant to appear to have a thousand leaves. Just as the millipede does not have a thousand legs, yarrow does not have a thousand leaves. Each leaf may reach a length of ten inches, though half that length is more usual. The leaves are deep green, and if one is chewed, a bitter aftertaste is left in the mouth. If the foliage is bruised, it is found to be strong-scented. Cattle avoid eating these leaves.

Achillea is another member of the Compositae, and its heads of flowers are produced in dense, flat-topped compound corymbs. Individual heads of flowers are relatively small if compared with another composite such as the daisy, and each head is composed of both ray and disc flowers. The small petals may be white or pinkish.

Yarrow's creeping runners must be dug out. If you try to fight yarrow chemically, you will find it moderately resistant to MCPA, 2,4-D, mecoprop, and even MCPA with dicamba. A number of applications may be needed.

Yarrow [*Achillea millefolium*]. **A.** Flowering stalk from rosette during second year's growth. **B.** Bipinnately dissected feathery leaves. **C.** Small heads of white flowers. **D.** Single head showing ray and disc flowers. **E.** Single disc flower. **F.** Single ray flower. The power of magnification is denoted by X.

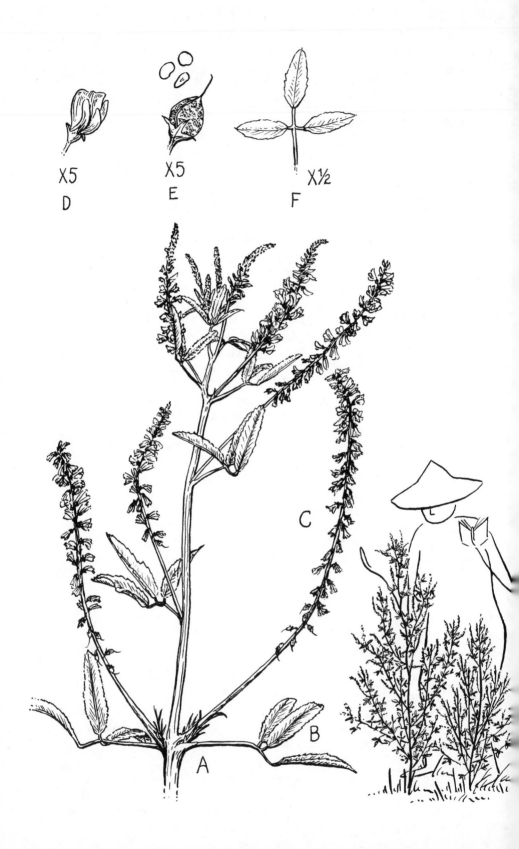

X5
D

X5
E

X½
F

A

B

C

Melilotus alba Desr.

COMMON NAMES White sweet clover, white melilot, honey clover, tree clover, honeylotus, cabul clover, bokhara clover, king's clover, hart's clover.

SOME FACTS Introduced from Eurasia; biennial; reproduces from seeds.

BLOOMS June to October.

RANGE Widespread throughout the United States.

HABITAT Sandy or gravelly fields, roadsides, waste places.

The seeds of both white sweet clover and its close relative yellow sweet clover (*Melilotus officinalis* Lam.) are popular with bee keepers, who grow these plants for their bees to visit. The yellow sweet clover is a bit more popular because it blooms somewhat earlier than the white and stops blooming a bit later. Both plants are used as forage plants, in which instance they are beneficial in two ways: they add nitrates to the soil because they have nitrogen-fixing bacteria within the nodules on their roots, and the shoot system is good for animals to eat if taken before flowering (after flowering the shoots become weedy).

Both clovers are popular fodder plants, but again, they should be harvested before flowering. Moldy melilot fodder may be poisonous, inducing internal bleeding because of the presence of the anticoagulant dicoumarin.

King's clover gets to be three to eight feet high, while the yellow melilot usually reaches only two to four feet in height. Hart's clover has a smooth, slender, much-branched stem. The leaves of both species are pinnately trifoliate, each leaflet being oblong to elliptical with very fine teeth. At the petiolar base are two stipules, both shorter than the leaf.

The typical leguminous flowers are produced on one side of a thin, spikelike raceme appearing in the axils of the upper leaves. The white quarter-inch-long blossoms are more fragrant than their yellow cousin's. Both produce small ovoid, wrinkled pods containing a few very long lived seeds. Those of the white clover have a coarse network of ridges while those of the red clover are cross ridged.

Do not allow white sweet clover (or yellow) to seed. Pull hard and try to get the thick fleshy root as well.

White sweet clover [*Melilotus alba*]. A. Flowering shoot. B. Single leaf of three leaflets. C. Flowering stalk with papilionaceous flowers. D. Single flower enlarged. E. Single fruit with seeds. F. A single leaf full face. The power of magnification is denoted by X.

A X1
B
C
D
E
F
G
X4

Pink- or Red-Flowered Species

Saponaria officinalis L.

COMMON NAMES Soapwort, bouncing bet, fuller's herb, scourwort, wood
phlox, mockgilly flower, hedgepink, wild sweet william, wild phlox.
SOME FACTS Introduced from Europe; perennial, reproduces by seeds and
rhizomes.
BLOOMS July to October.
RANGE Temperate North America.
HABITAT Along railroad embankments, roadsides, empty lots, waste places.

With all those common names it's surprising no one has called
Saponaria officinalis railroad weed; it was along railroad tracks that as a
youth I first saw bouncing bet; though I had not the slightest idea what
it was, and worse, I did not care!

Colonial maidens dared not use their coarse lye soaps on their delicate
woolens or silks, so soapwort stems and leaves were gathered, broken in
water, and in the resulting sudsy water fine materials were not only cleaned
but given a special luster. Even the flowers may be bruised in water and
their saponin content put to use.

One underground rhizome will give rise to many stems, thus accounting
for the great number of individual plants usually found growing together.
Individual plants of bouncing bet may grow two or three feet high, and
are smooth with swollen joints. The stems may remain unbranched or may
branch sparingly.

Wild phlox has opposite, simple leaves that lack petioles and are ovate-
lanceolate in shape with smooth margins. Close inspection reveals what
appears to be parallel venation but is actually palmate venation.

The lovely pink five-petalled blossoms (the petals are fused) make
their appearance in dense terminal clusters, though others may appear in
the axils of opposite leaves lower down on the stem. The five sepals are
seen to be fused into a long tube at the base of the flower.

Soapwort should be poisonous, as the soap-forming principals are poisonous, but few reports of poisoning have been verified. Animals generally
seem to find the weed distasteful.

The plant has been known since the days of Dioscorides and has been
used to treat venereal disease, though this use has long been abandoned.
Saponaria causes trouble in the grain crop grown in our Northwest.

Dig out plants of bouncing bet. Spraying with sodium chlorate will
also help. Hot brine is said to help, too, but the soil will be useless until
leaching has taken place.

Soapwort [*Saponaria officinalis*]. A. Flowering shoot. B. Opposite three-veined leaves.
C. Five-parted flowers showing petals separate at top and fused at bottom. D. Pistil
(right), stamens (left). E. Mature fruit, a capsule. F. Center of capsule with the many
tiny seeds attached. G. Single seed enlarged. The power of magnification is denoted by X.

Rumex acetosella L.

COMMON NAMES   Field sorrel, red sorrel, sourweed, sour leek, little vinegar plant, cow sorrel, horse sorrel, sour dock.

SOME FACTS   Introduced from Europe; perennial; reproduces by seeds and creeping rootstocks.

RANGE   Throughout the United States and Canada.

BLOOMS   May to October.

HABITAT   Old cut fields, pastures, meadows, highway embankments.

Sourweed will be seen spreading rapidly wherever the ground has become acid, and where nitrates have been leached from the soil. Sorrel means sour.

Rumex acetosella is much smaller than its first cousin *R. crispus* (p. 123). Red sorrel's stem never gets larger than a foot high, and is slightly bent near the base. Furthermore, it produces rapidly spreading rhizomes, which, fortunately, are not produced by the larger *Rumex*.

The leaves of this weed are interesting in that they are halberd-shaped (hastate) and are spicy in their sourness when bitten. Hikers, versed in nature lore, will eat some of the leaves as a thirst quencher. Sour dock's leaves are also used in salads or soups (sorrel soup).

The sourness arises from the presence of crystals of calcium oxalate, a dangerous poison that is deposited in the leaves; however, a few leaves of red sorrel will not harm anyone, and if used to make sorrel soup, the boiling will reduce the oxalate content drastically. Our food plant rhubarb (*Rheum raponticum*), also a relative to *Rumex*, has leaves loaded with calcium oxalate that are very dangerous to include in the pie. The petioles that are usually used for rhubarb pie have much less oxalate in them.

The female flowers are tipped with three tiny bright red feathery stigmas, and the whole flower (female) is red at maturity (hence the common name red sorrel). When the stalks with their mature female flowers are all up, the field or roadside takes on a red glow that is not unattractive. The inflorescence takes up approximately half the stem of this relatively short plant.

Kalm tells us in the eighteenth century, "Blue is dyed with indigo, but to get black, the leaves of the common field sorrel (*Rumex acetosella*) are boiled with the material to be dyed, which is then dried and boiled again with logwood and copperas. The black thus produced is said to be very durable."

Should the plant take root in your soil, either pull it up at the first opportunity or use MCPA salt or 2,4-D amine, and you should rid yourself of this acid-loving plant. Adding nitrates and lime to the soil will also eliminate it.

Sheep sorrel [*Rumex acetosella*]. A. Flowering stalk of sorrel. B. Halberd-shaped leaves with eared bases. C. Spikes of tiny flowerets. D. A mature fruit.

Trifolium pratense L.

COMMON NAMES   Purple clover, red clover.
SOME FACTS   Introduced from Europe; perennial; reproduces by seeds.
BLOOMS   May to August.
RANGE   Throughout temperate North America.
HABITAT   Lawns, meadows, pastures.

In addition to purple clover (and other clovers), the legume family includes tremendously economically important plants such as alfalfa, soy beans, peas, beans of all kinds, peanuts, and the lovely garden sweet pea. Most species of leguminous plants have small tubercules on their roots in which dwell various species of a bacterium known as *Rhizobium*. This bacterium is capable of fixing atmospheric nitrogen-forming nitrates, which can then be taken into plants by absorption through their roots and built into amino acids, themselves the building blocks from which proteins are formed. Therefore, the root tubercules of the legumes are of vast agricultural significance. The stems of red clover may become two feet long, and may either grow erect or rest on the ground. These stems are covered with soft hairs.

Purple clover's leaves are divided into the three leaflets expected in clovers and have V-shaped whitish areas across the middle of each leaflet, the point of the V aimed at the point of the leaf. Lower down on the stem the leaves are on petioles, while higher up they tend to be sessile.

The "flower" of clover is a cluster of many flowers, globose in shape, and red or purple in color, and are terminal upon the stems that bear them. Each individual flower of the cluster is typical of the legume-type flower, with a banner upper petal, wings, and a keel.

After pollination by bumblebees, each individual blossom is lowered from its former relatively upright position in the cluster and turns brown as seed set occurs.

Red clover is frequently planted for forage. Digestion of large quantities of red clover makes horses oversalivate (called the slobbers) and has been known to impair or destroy the vision of cattle. In spite of this Gerard noted in 1597 that "oxen and other cattell do feed on the herb as also calves and young lambs. The flours are acceptable to bees. Pliny writeth and setteth it downe for certaine that the leaves hereof do tremble and stand right up against the coming of a storme or tempest."

Red clover is resistant to MCPA and 2,4-D amine and ester, and it is moderately resistant to mecoprop salt. Repeated application of the latter may be successful.

Purple clover [*Trifolium pratense*] and white clover [*T. repens*]. **A.** Purple clover flowering stalk and leaves, with three leaflets, a white V on each leaflet. **B.** Head of papilionaceous flowers. **C.** Individual flower. **D.** Nodules on root containing nitrogen-fixing bacteria. **E.** Leaves and flower heads of white clover. **F.** Turned-down pollinated flowers. The power of magnification is denoted by X.

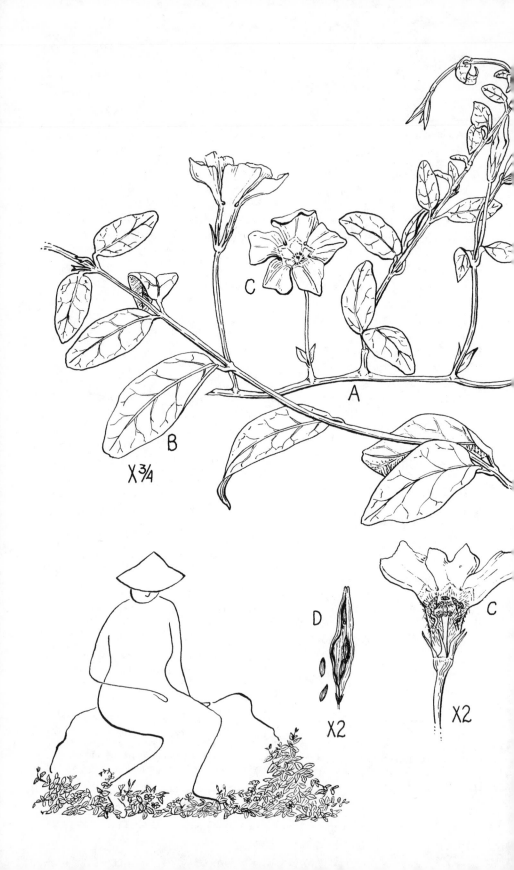

A

B

X ¾

C

C

X2

D

X2

Vinca minor L.

COMMON NAMES Periwinkle, myrtle, small periwinkle.
SOME FACTS Perennial; arrived from Europe as a garden plant; propagates through runners.
BLOOMS February to June.
RANGE Widespread in northeastern United States; also on Pacific Coast.
HABITAT Moist rich soil; bordering lawns, gardens, roadsides, cemeteries.

The lovely blue funnel-form flowers of periwinkle make it difficult to think of this plant as a weed, but it spreads so fast by means of its runners that it can rapidly become weedy and take over areas where it is not wanted. Perhaps this is because it has forty-six chromosomes (precisely the chromosome number of the human being). We too seem rapidly to take over areas where we are not entirely wanted.

This relative of the dogbanes (p. 191) has creeping stems that root at the nodes. Along it are dispersed the simple entire oblong to ovate leaves that are firm to the touch, glossy, and evergreen. Its solitary blue flowers with their five petals forming a funnel-shaped corolla are within the axils of the leaves and have truncated lobes on their edges.

While periwinkle produces fruits, two short cylindrical follicles, reproduction by seed is far, far rarer than by runners.

John Gerard in 1598 reminded his readers that "the leaves boiled in wine and drunken, stoppeth laske and bloodie flux; it likewise stoppeth the inordinate course of the monethy sicknesse."

I have yet to encounter even one human being who ever wanted to suffer from the "monethy sicknesse." Have you?

Pull up with gloved hand or use simazine, which has some effect on *Vinca*.

Periwinkle [*Vinca minor*]. **A.** A portion of the prostrate stem. **B.** One of two opposite, pinnately veined leaves, which are glossy and dark green. **C.** The sizeable five-petalled blue flower full face and in side view. **D.** Double follicle opening and shedding seeds. The power of magnification is denoted by X.

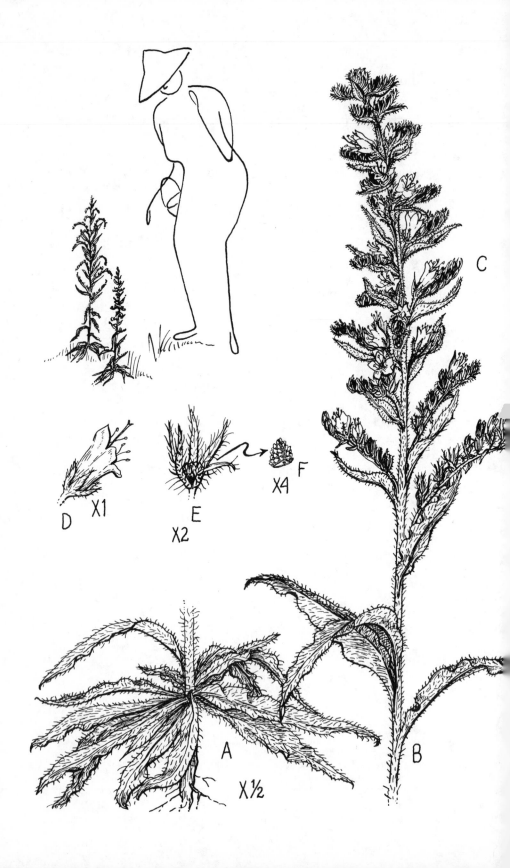

A X½

B

C

D X1

E X2

F X4

Echium vulgare L.

COMMON NAMES Blueweed, viper's bugloss, blue devil, blue thistle, viper's herb, snake flower.
SOME FACTS Introduced from Europe; biennial; propagates by seeds.
BLOOMS June to September.
RANGE New Brunswick to Ontario, south to Nebraska and Georgia.
HABITAT Meadows, pastures, waste places, poorly drained slopes, roadsides.

In ancient times this plant was thought to discourage serpents, and its root, dissolved in wine, was thought to be good for snakebites; this has not been confirmed by modern science.

Blueweed, known by so many common names, is rooted to the soil by a sturdy, dark, long taproot. During the first of its two years of growth a rosette of entire oblong to lanceolate leaves makes its appearance. Some of the leaves may grow to six inches in length; all are covered with bristly hairs on both surfaces and are dark green.

Blueweed's long taproot seems happiest on gravelly shaley slopes, and under these conditions blueweed may dominate the landscape. During the second year of growth a bristle-covered stalk grows from the center of the rosette and may reach one to three feet in height. The bristles are so numerous that the stalk and its leaves look gray-green from a short distance. The bristles, which begin as stiff hairs, harden into outright prickles. Each of them arises in a red tubercule that specks the stem.

The lovely flowers of blue thistle are produced in curved racemes (sometimes called cymes), which arise in the upper axils. A five-lobed, irregular, funnel-shaped corolla is pink in the bud, and bright blue at maturity. (The color change is due to a shift from acidity to basicity of the sap in the vacuoles of the epidermis of the petals.) Since the stamens remain red, they stand out against the bright blue of the mature blossom and the pollen they shed is among the smallest known of the flowering plants, only one to fourteen microns (1 micron = 1/1000 mm).

The open blossoms are not fragrant to us, but the bumblebees visiting them apparently are aware of a scent. After pollination, four small wrinkled nutlets result and are very long lived. They are frequently found as constituents of other batches of seeds.

The Australian cousin of *Echium vulgare*, *E. plantagineum*, is so disliked and unwanted that the government has outlawed the plant! *E. plantagineum* contains an alkaloid that is a strong liver poison.

Hand pulling of blue thistle is difficult because of the pricklelike hairs, but with a thick pair of gloves you should be able to get at the taproot, which yields to force. The plant is moderately resistant to MCPA and 2,4-D amine. To MCPA plus it is moderately sensitive.

Blueweed [*Echium vulgare*]. A. Rosette of hairy linear leaves of first year's growth. B. Flowering stalk of second year's growth, also hairy. C. Individual flowers, pink before opening, blue afterward. D. Individual flower. E. Mature fruit within sepals. F. Fruit enlarged.

X2 D

X4

E X1

B

C

A

X1

Cichorium intybus L.

COMMON NAMES Chicory, cornflower, succory, ragged sailors, witloof, wild endive, coffeeweed, blue daisy.

SOME FACTS Introduced from Europe; perennial; propagates by seeds.

BLOOMS July to October.

RANGE Nova Scotia to Florida, west to the plains.

HABITAT Vacant lots, fields, roadsides.

Cichorium intybus, chicory, was cultivated for its usefulness in Egypt and the Middle East. *Cichorium* is the Latin version of the name used formerly for a species of this plant in Arabia, and *intybus* is a corruption of *hendibeh*, another Eastern name for chicory.

The plant's leaves make a fine salad base or potherb, though one should choose the young leaves of the rosette. Succory is a first cousin of endive; its taproots are tasty cut up and placed in salads, or dried, powdered, and used as a coffee adulterant. There are those who would insist that chicory is not an adulterant of coffee but an enhancer. Coffee served in the South often contains chicory. Indeed, chicory is sometimes used alone as a coffee substitute. The root of *Taraxacum officinale* (the dandelion, p. 135) is used in precisely this way and it is interesting to note that chicory and dandelion are first cousins.

Silver-dollar-sized bright blue flowers are produced in clusters of one to four each on very short pedicels. Only one flower of each cluster blooms at a time—a very short time, opening early in the morning and closing by noon on bright days, but remaining open for a longer period on darker days. The blue daisy is not exactly similar to the real daisy, for the real one has both ray and disc flowers, while the blue version has only ray flowers (as do the dandelions). Withering of the flower is independent of the pollination of the flower.

Ragged sailors' blossoms are distributed along a two- to five-foot hollow stem that is sparsely branched. The leaves are roundish, four to eight inches long, and alternately distributed along the stem; those lower on the stem are smaller than those of the rosette at the base of the stem, and are clasping and eared. Those of the rosette are spatulate, narrowing into marginal petioles. If you look at the underside of cornflower's rosette leaf, you will see stiff hairs along its midrib.

If the soil is moist, merely pulling out by hand should rid you of wild endive (as Peter Kalm calls this pretty weed), but be sure to get the taproot. Chicory is also moderately susceptible to MCPA salt (3 ounces per acre) and to 2,4-D amine (2 ounces per acre).

Chicory [*Cichorium intybus*]. A. Flowering stalk with (B) dandelionlike leaves and (c) silver-dollar-sized light blue heads. D. Single ray flowers from head, which contains only ray flowers. A seed is to the right of the ray flower. E. The thickened taproot, which can be heated and ground for a coffee substitute.

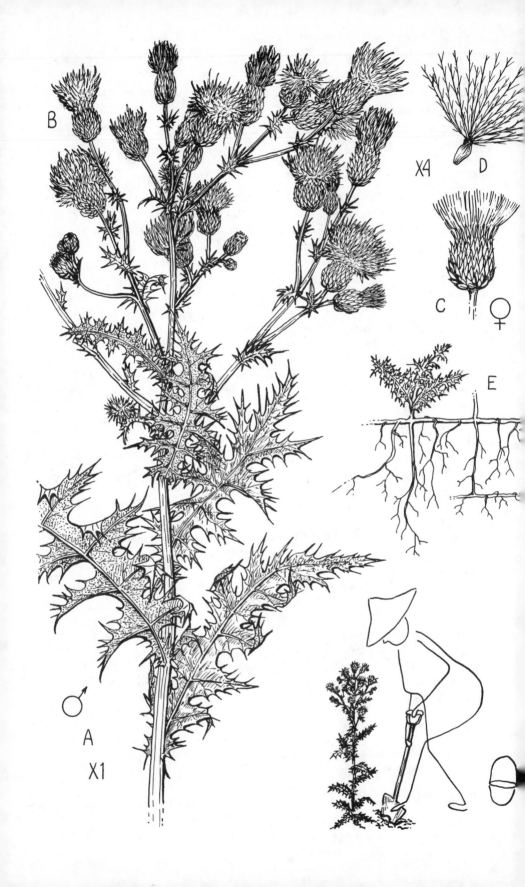

B

X4   D

C   ♀

E

♂
A
X1

Purple-Flowered Species

Cirsium arvense Scap.

COMMON NAMES Canada thistle, creeping thistle, green thistle.

SOME FACTS Introduced from Eurasia; perennial; reproduces by rhizomes, rarely by seeds.

BLOOMS July to October.

RANGE Common to southern Canada and northern United States, as far south as Virginia and northern California.

HABITAT Pastures, meadows, cultivated fields, waste places.

Listen to the generally reserved English on *Cirsium arvense*: "The Minister has power under the Act [of Parliament] to serve notice on an occupier of land requiring him, within the time specified in the notice, to take such action as may be necessary to prevent these injurious weeds from spreading [one of them is *Cirsium arvense*]. An occupier who unreasonably fails to comply with the requirements of such a notice is liable, on summary conviction, to a fine not exceeding £75, or for a second or subsequent offence, not exceeding £150. . . ." The British aren't the only ones upset about this plant; almost every state in the Union has outlawed it!

Canada thistle or creeping thistle is a very difficult plant to get rid of once it has taken hold because it produces rapidly growing rhizomes, which send up above-ground shoots at short intervals.

The erect, grooved stem may be one to four feet high and is woody. Its heavily armed leaves are three to six inches long, sessile irregularly pinnatifid. Each is toothed with hard, white needlelike spines that point this way and that.

Cirsium, the small-flowered thistle, is a composite, and therefore produces its flowers in heads: these are terminal or axial clusters, and pink or lavender-purple in color; all are disc flowers. The plants are single-sexed, which is a blessing. If only one sex has sprung up in your garden, your problem will be to fight the impossibly fast-growing rhizome; you will not have to worry about the seeds.

Mature seeds must wait until the plant dies and the heads are dropped to be freed, in spite of the fact that a rich supply of down is associated with the seeds in the head. When the head matures, the down flies away unattached to anything. Seedlings are rarely encountered.

Gerard, commenting on this thistle (called by him *Acanthium album*), writes of the "vertues": "Dioscorides saith, That the leaves and roots hereof are a remedy for those that have their bodies drawne backwards." God keep that from us!

Frequent mowing will prevent the growth of *Cirsium*. Both amitrole and 2,4-D may be used to eradicate this garden menace.

Canadian thistle [*Cirsium arvense*]. A. Shoot showing the dangerously armed, dandelion-like leaves. B. The heads of purple flowers. C. Single head, natural size. D. An enlarged fruit with its parachute attached. E. A section of the rhizome showing how this plant can travel underground and maintain itself. The power of magnification is denoted by X.

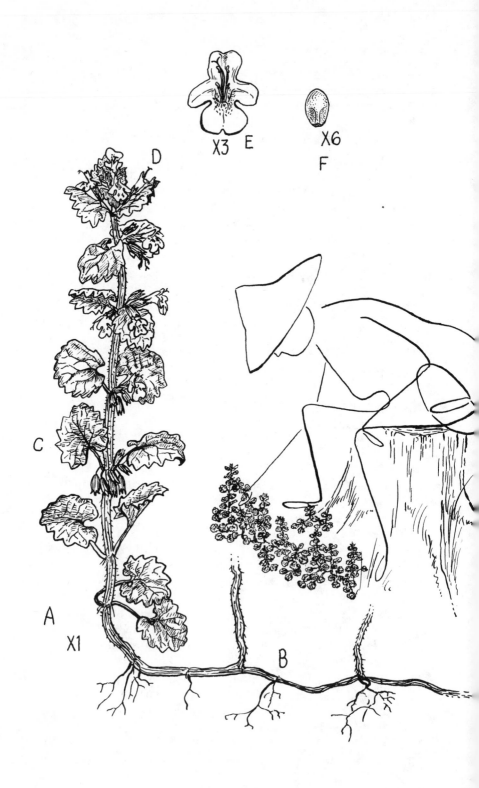

X3 E

X6
F

D

C

A
X1

B

Glechoma hederacea L.

COMMON NAMES Ground ivy, gill-over-the-ground, field balm, creeping charlie, cat's-foot, runaway-robin, ale-hoof, and gillale.

SOME FACTS Perennial; reproduces by rootstocks and by seeds; introduced from Eurasia.

BLOOMS April to July.

RANGE Widespread throughout northern United States and southern Canada; also in American South.

HABITAT Damp, shaded ground.

The appearance of the word "ale" in two of the common names given above gives a strong hint of one of the uses of this very common lawn weed. It was once used in place of hops for the flavoring and preparing of homemade ale. The gill in the common names points to the same use, for gill comes from *giller*, which is French for ferment.

On its creeping, sometimes eighteen-inch-long stem, ground ivy sports small orbicular leaves with crenate edges. That this plant is a member of the mint family is given away by two things: its square stems, and its two-lipped flowers, which are bilaterally symmetrical.

Clusters of these pale purple flowers appear in the axis of the leaves and give rise to nutlets that are brown but have one white spot.

The late Euell Gibbons reminded us in *Stalking the Healthful Herb* that this plant can be used as a bitter tonic and nutritious tea that is an excellent remedy for a stubborn cough, and that it is loaded with vitamin C.

Spraying carefully with sodium chlorate should rid you of ground ivy.

Ground ivy [*Glechoma hederacea*]. **A.** Habit sketch of plant. **B.** Roots springing from the rhizomes. **C.** Orbicular, crenate leaves, opposite and with clusters of flowers in the axils. Stem is squared. **D.** Small two-lipped flower. **E.** The bilaterally symmetrical flower. **F.** A much enlarged nutlet with a white spot. The power of magnification is denoted by X.

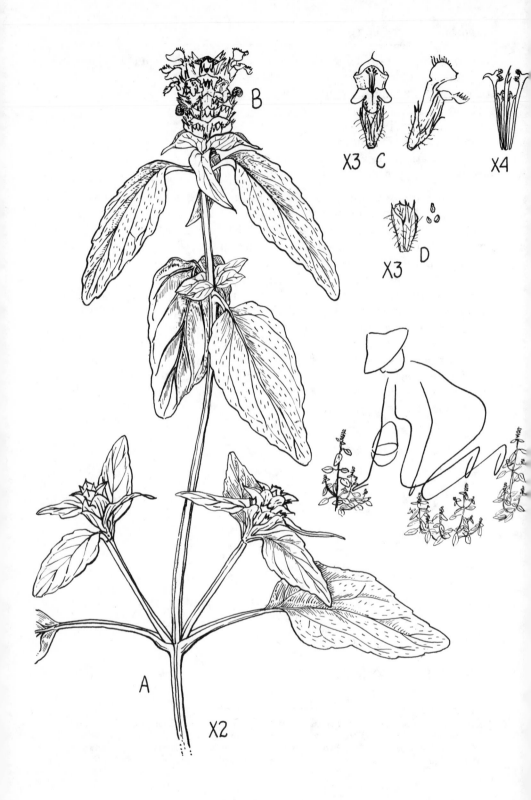

B

X3 C

X4

X3 D

A

X2

Prunella vulgaris L.

COMMON NAMES Heal-all, selfheal, carpenter weed, prunella, sicklewort, heart-of-the-earth.

SOME FACTS Native of the United States and most of the rest of the world; perennial; reproduces by seeds.

BLOOMS May to October.

RANGE Widespread throughout North America.

HABITAT Lawns, pastures, waste places, meadows.

Heal-all, which actually heals nothing, is called *Prunella* in error; just as *Verbascum* is a corruption of *Barbascum* (p. 149), *Prunella* is a corruption of *Brunella*, from the German word for quinsy, a type of sore throat (*die Bräune*) for which selfheal used to be a treatment. Its bruised leaves were wrapped around the throat. Our own Indians used selfheal to cure dysentery in babies. These uses have not withstood the rigorous testing of modern medicine.

This is one of the very few weeds that almost every gardener in the world may find growing on his land, usually in poorly drained soils. Sicklewort is native to the United States, and to Europe and Asia as well, a really ubiquitous weed. If you happen to be in England, you might check on Gerard's statement that Prunella "grows in Essex neere Heningham castle" and that those specimens were white flowered.

This plant is not entirely without charm. Its blue-violet (sometimes white or pink) tubular sessile flowers are two-lipped, as are members of the mint family; the upper lip is arched and the lower lip is spreading and three-lobed. These blooms occur within a thick spike within the axils of bractlike leaves.

Depending on growth conditions, carpenter weed may be two inches to one foot high. In the sun the plant is short and darker green than in the shade. Constant mowing of a lawn infected with heal-all will cause horizontal spreading of the stems, which, since they can root at their nodes, will only spread the plant more rapidly.

Its leaves are oppositely disposed along the square stems so typical of mints, and are oblong-ovate, broader at their bases.

Prunella may be dug out if the patch is small. If it is large, use 2,4-D amine, to which it is moderately sensitive. An 8 percent solution of iron sulphate should kill selfheal and spare your lawn grass.

Heal-all [*Prunella vulgaris*]. **A.** Square-stemmed shoot with opposite leaves. **B.** Spike of flowers at top of stem. They may also be seen emerging from the axils of leaves. **C.** Full face and side view of typical two-lipped bilaterally symmetrical corolla of a flower belonging to the mints. **D.** Fruit with tiny seeds. The power of magnification is denoted by X.

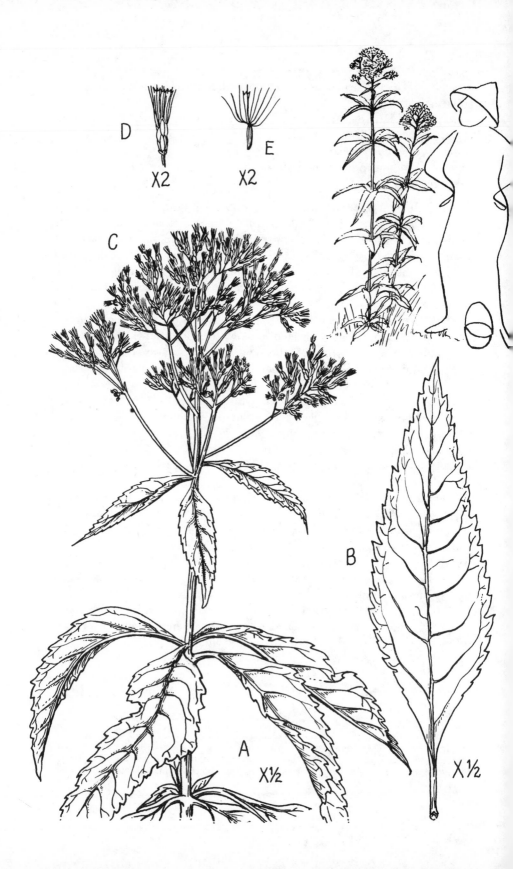

Eupatorium purpureum L.

COMMON NAMES   Joe-pye weed, purple boneset, tall boneset, trumpetweed, feverweed, queen-of-the-meadow, gravel-root, kidney-root.

SOME FACTS   Native to the United States; perennial; propagates by seeds.

BLOOMS   August to September.

RANGE   New Brunswick to Manitoba, southward to Florida and Texas.

HABITAT   Damp meadows, borders of moist woods, sides of streams, along ditches.

There are a number of species of the genus *Eupatorium*, and one of them was used by Mithridates Eupator, King of Pontus, as an antidote for a certain poison. In finding the antidote he gave his name (which would otherwise have been forgotten) to the plant. *Eupatorium* is a composite and a close relative of the garden plant *Ageratum*.

The commonest common name, Joe-pye weed, comes directly from the name of an American Indian of our own Northeast. Joe-pye was an herbal doctor who specialized in tonics, decoctions, and medicines concocted from parts of this plant.

One of the taller weeds, this one may tower above you; its three- to ten-feet smooth, ridged stems are capped with a great flat-topped corymb of purple heads. These stems are usually speckled with dark purple. Its large oblong-ovate to lanceolate leaves are arranged in whorls of three to six leaves each. The leaves are smooth above but hairs may be found along the veins of the underside.

Each purple head of the large flat corymb contains only tubular disc flowers. Unlike daisy or sunflower, no ray flowers are present in the head of purple boneset.

Any trumpetweed that arrives on your land may be removed by hand pulling.

Joe-pye weed [*Eupatorium purpureum*]. A. Shoot showing leaves arranged at node. B. Large single leaf of this tall plant. C. Individual flowers in loose clusters. D. Single disc flower. E. Mature achene with parachute attached. The power of magnification is denoted by X.

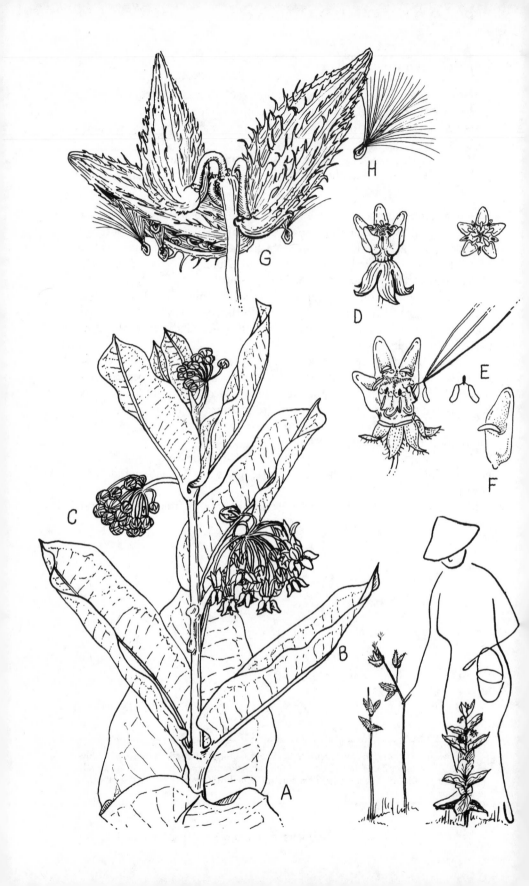

Asclepias syriaca L.

COMMON NAMES Milkweed, silkweed, cottonweed, silky swallowwort, Virginia silk, wild cotton.

SOME FACTS Native; perennial; reproduces by seeds and by rhizome.

BLOOMS June to August.

RANGE Common in eastern North America, west to Iowa and Kansas.

HABITAT Fields, pastures, waste places, roadsides; often on rich gravelly or sandy loam.

Asclepias was the god of medicine to the Greeks (he was called Aesculapius by the Romans). His name was given the plant because it was once used in a number of medicinal concoctions; it is still used in a formula supposed to help the asthmatic. Animals that ingest its leaves, and a resinoid contained in them, show a number of frightening symptoms—profound depression and weakness, spasms, labored respiration, elevation of the temperature with dilation of the pupils (i.e., all the symptoms a man in love endures), and then, finally, a comatose state ending in death.

Interestingly, despite these frightening facts, the young shoots make a delightful asparagus substitute if cooked. Boiling in water removes the dangerous principle. Even the older leaves may be cooked (after being rinsed a number of times like spinach), and they taste very good. The American Indians were fond of cooked unopened flower buds of this plant, and the very young pods are cooked and eaten very much the way okra is.

Few have not seen the two- to five-inch warty pods opening in the fall and shedding their cargoes of fine, silken-haired flat brown seeds. The hair of these plumose air-borne seeds is used to stuff pillows today as it was during the eighteenth century. Let us listen to Kalm: "The pods of this plant when ripe contain a kind of wool, which encloses the seed and resembles cotton, whence the plant has gotten its French name [*le cotonnier*]. The poor collect it and with it fill their beds, especially their children's, instead of feathers."

The five-petalled, dull purple to pink flowers are borne in umbels. Each stamen bears an appendage on the side away from the pistil, which enfolds the antherlike hood. Mature blossoms are pollinated by the gorgeous monarch butterfly (also called the milkweed butterfly), *Danaus plexipous*.

If the tops are cut from the ground level before pod opening, and efforts are made to cut and starve the rhizome, the growth of *Asclepius syriaca* can be terminated. Repeated treatment with sodium chlorate should also destroy this weed.

Milkweed [*Asclepias syriaca*]. **A.** Shoot showing opposite, oblong, leathery leaves. **B.** Leaves will bleed milk if damaged. **C.** Umbels of flowers still to open. **D.** Single five-parted flower. **E.** Instrument removing pollen sacs, which insect, usually a butterfly, removes from one plant and deposits on the flowers of another. Also shown are pollen sacs looking something like saddlebags. **F.** Where pollen sacs land. **G.** Large mature fruits, which open along one suture (follicles). **H.** Individual parachuted seeds escaping.

A

B

C
X1

D X3

X¾

Arctium lappa L.

COMMON NAMES   Great burdock, beggar's buttons, cockle-button, clotbur.
SOME FACTS   Introduced from Europe; biennial; reproduces by seeds.
BLOOMS   July to October.
RANGE   Widespread in the northeastern and north central states, and here and there on the Pacific Coast.
HABITAT   Roadsides, waste places, fence rows, neglected farmyards.

If you know what rhubarb looks like but have never really looked closely at its huge leaves, you might easily mistake the first year's growth of great burdock for the rhubarb plant; however, the resemblance is purely superficial. The strikingly large leaves of the first year's rosette may grow to over one foot long; their lower surfaces are light green and woolly, while their upper surfaces are darker and smooth. In this species the petioles are solid. The first year's leaves arise from a short stem attached to a large robust, deeply buried taproot that may be a foot long and three inches wide.

You may, indeed, have unknowingly eaten this root; the sliced roots of a domestic variety (called gobo) are used in sukiyaki. In Hawaii gobo is thought to endow the eater with strength of body, and since it is also thought to have aphrodisiacal qualities, with other kinds of strengths as well. Gobo is collected only during its first year's growing period.

During the second growing season of its two-year life clotbur sends up a tall, flowering stem that is thick, hollow, and grooved. It branches and produces alternate leaves, which are smaller in size than the leaves of the earlier rosette, but of the same general shape. In the axils of the upper leaves on this rough-hairy stem are borne the clusters of floral heads. *Arctium* is also a composite.

Each nearly globular head may be more than an inch across and is usually on a longish peduncle. Individual flowers are pink-purple and tubular—there are no ray flowers—and have all their floral organs, including purple anthers and white stigmas on the pistil. Each bract of the involucre is hooked.

The burs attach themselves with great ease to animal fur as well as to the clothes of passing human beings. Elementary school boys are often guilty of throwing the burs into elementary school girls' long tresses, or were in the days when life was still simple and sweet at that age.

One plant may produce four hundred thousand seeds, so it would behoove the owner of this easily detected weed to get rid of it during its first year's growth. Cut off the rosette and then dig out as much of the powerful taproot as you can get at. Pouring a concentrated brine solution into the cut taproot will not hinder your aim. The plant is moderately sensitive to MCPA salt and to 2,4-D amine.

Great burdock [*Arctium lappa*]. **A.** Large leaf from first year's rosette. **B.** Second year's flowering stalk showing smaller leaves and the mature fruits armed with barbs to attach them to fur or clothes. **C.** Single head of flowers. **D.** Single disc flower. The power of magnification is denoted by X.

A X½

B

C

D X2

E

F X2 X

Desmodium canadense DC.

COMMON NAMES Tick trefoil, showy tick trefoil, stick-tight.
SOME FACTS Native to the United States; perennial; reproduces by seeds.
BLOOMS August to September.
RANGE Generally in eastern North America, south to the Carolinas.
HABITAT Borders of woodlands and streams, old fields; often on gravelly
soils.

There are many species of the genus *Desmodium* but all are recognizable by their pods. The pod of *D. canadense,* showy tick trefoil, is about
an inch long and is slightly curved. It contains three to five seeds and is
covered with tiny hooked hairs that cause it to stick tightly to your clothing if you brush against the mature pods during one of your nature walks.
The pods usually break into sections, each one a small oval or triangle
containing one seed, when you brush against it. Since among the species of
Desmodium are tall plants, prostrate ones, and middle-sized ones, a walk
in the woods in late September can leave you covered from head to toe
with the little stick-tights.

Showy tick trefoil produces typically leguminaceous flowers in racemes
at the top of the branched stem. The racemes are densely flowered and
contain ovate-lanceolate bracts. Most blossoms are bluish-purple, though
some are nearly white. Bluish-purple flowers are rose-purple before opening, gradually assuming a darker hue.

The plant's erect, stout stem is branched above, is ridged and grooved
and very hairy. Each leaflet of the trifoliate leaves that are borne on the
stem is oblong-ovate with a nearly smooth upper surface, though the lower
surfaces are covered with finely appressed hairs. Each leaflet has numerous
nearly straight veins. Leaves produced on the upper branches are nearly
sessile.

Put on gloves and pull up before the pods appear. If they have already
appeared, wear a bathing suit to offer as small a clinging surface area to
the pods as possible.

Tick trefoil [*Desmodium canadense*]. A. Flowering shoot. B. Palmately-compound leaves
with three leaflets (hence, trefoil). C. Many flowering inflorescences and mature fruits
can be seen here and there. D. Single flower with typical papilionaceous petal arrangement. E. Single mature fruit (legume that breaks into sections), covered with hooks
that attach it (in seconds) to the passerby. F. Single seed. The power of magnification
is denoted by X.

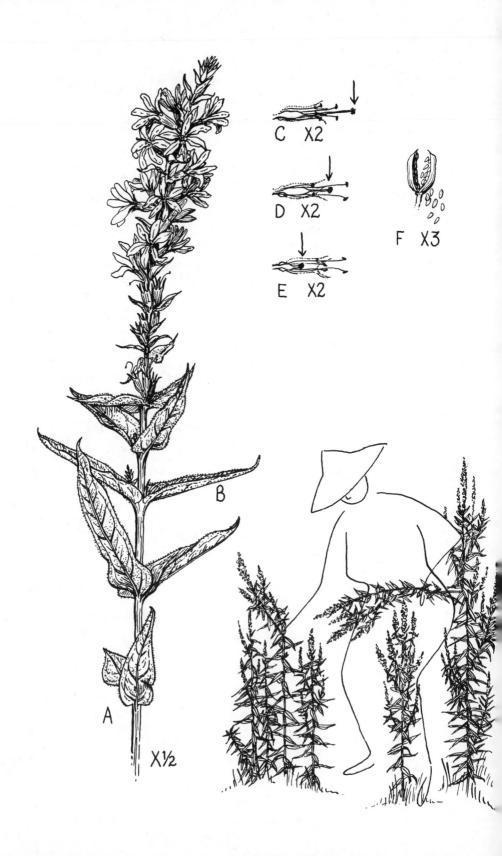

A

B

C X2

D X2

E X2

F X3

X½

Lythrum salicaria L.

COMMON NAMES Purple loosestrife, willow herb, spiked loosestrife, bouquet-violet.
SOME FACTS Perennial; reproduces by seeds; arrived from Europe.
BLOOMS June to September.
RANGE Northeastern United States to Virginia and Missouri; also on Pacific Coast.
HABITAT Marshes, along shores and ponds, wet soil.

Some readers may question the inclusion of purple loosestrife among the weedy species of the flowering plants—especially someone who has done what I have done: pulled off the road just to stop and stare at the flaming beauty of a pond in early autumn when it seems one mass of purple. Yet, this spectacle suggests a reason for inclusion. The very abundance in which the plant grows prevents other plants from having *Lebensraum*. Purple loosestrife literally chokes out native vegetation. And that is weediness. The definition of weed is not entirely based on disruption of the farm, garden, or lawn.

The tall stems of willow herb may sometimes reach four feet, and along this stem are borne opposite, sessile, lanceolate to nearly oblong leaves that taper to a point and may have white hairs on them. Sometimes the leaves are whorled, usually with three leaves at a node.

Its stunning blooms are produced within the axils of the leaves, leaves that are willowlike and give the plant its Latin specific epithet, *salicaria*, and its common name, willow herb. Its generic name refers to the blood-like color of its flowers: *Lythrum* is derived from the Greek *lythron*, meaning blood.

Purple loosestrife usually stands with its feet in water or in very wet soil; thus great care should be used when trying to get rid of it with herbicides because these may endanger other water plants. It is moderately resistant to 2,4-D amine, but quite resistant to most other chemicals. You could try hand pulling, but be sure to wear your gloves . . . and your rubbers.

Purple loosestrife [*Lythrum salicaria*]. A. Shoot showing the sessile, opposite leaves. B. Each leaf is wide at the base, tapers to a point, and is lightly hairy. C, D, E. The three different lengths of pistils found in the blooms of the species. This condition is known as trimorphy (three-shaped): (C) where the pistil extends far past the stamens; (D) where the pistil is below the anther sacs of the emergent stamens; (E) where the pistil is short and the stigma down within the flower. All three aid in the pollination of this flower. F. An opening capsule with its small seeds. The power of magnification is denoted by X.

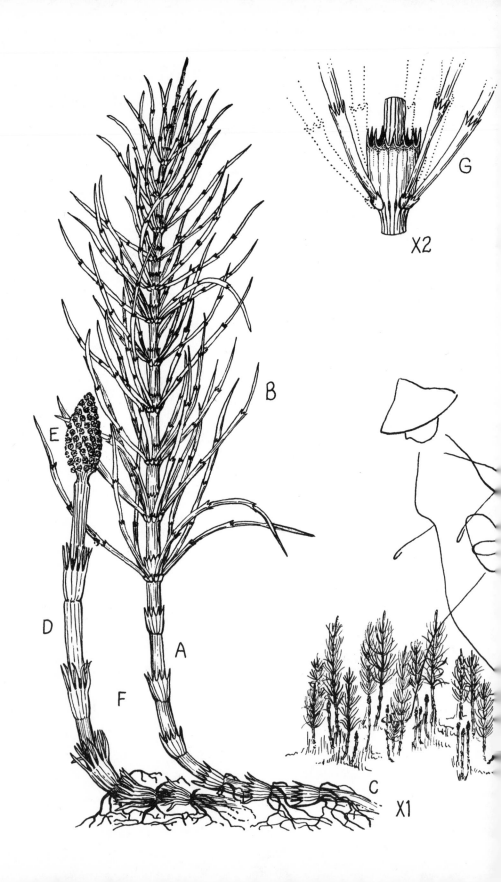

Equisetum arvense L.

COMMON NAMES Field horsetail, scouring-rush, horsetail fern, pinegrass, foxtail rush, bottle brush, horsepipes, pinetop.

SOME FACTS Native to United States and southern Canada; perennial; reproduces by spores (not seeds), creeping rootstocks, and tubers.

SPORULATION April to May.

RANGE Throughout the United States and southern Canada.

HABITAT Moist fields, meadows, moist road embankments.

The field horsetail is not a flowering plant, but is common enough—especially where it is not wanted—to have earned the name of weed. Instead of producing flowers and seeds, scouring rush produces a cone atop a jointed, yellowish, fluted stem; in the cone are formed thousands of small green four-winged spores. The cone before full maturity looks something like a papal tiara in overall shape.

After the spores have been shed, the fertile yellowing cone-bearing shoots die back to the jointed, perennial rhizomes, and then sterile green shoots come up from the rhizome. These shoots are eight inches to one foot in height, and are jointed and fluted, as were the fertile shoots; they are hollow between nodes and solid at the nodes, where sheaths composed of scalelike leaves that are fused at their bases are formed. The sheaths' tips appear as small "teeth" around the node.

What may unknowingly be interpreted as long, thin green leaves are actually branches, and they occur in whorls at the solid nodes of the main stem; each thin stem may also give rise to still thinner stems at each node, and thus a plumose plant that looks something like a horse's tail will be seen. Sterile shoots remain green and alive until autumn.

Horsetail apparently contains one or more thiamine-destroying principles, though which of the many potentially poisonous chemicals it is has not yet been determined. Horses are particularly susceptible to poisoning by horsetails and many have died of eating this plant. Administration of thiamine brings the animal back to health, unless it has already reached the last stages of this disease, known as equisetosis.

Maidens in the seventeenth century, lacking Brillo pads, used scouring rush to clean and polish metal dishes, especially to polish pewterware. The plant is still used today to give a fine finish in cabinetmaking.

The aerial shoots are quite sensitive to MCPA, MCPB, and 2,4-D, which should be applied when the shoot is at maximum height; however, the presence of deep rhizomes and their associated tubers make getting rid of *Equisetum* a bit of a task. Since it likes moist, sandy, or gravelly soil, creating better drainage may also do the trick.

Scouring-rush [*Equisetum arvense*]. A. Vegetative shoot (note jointing of stem). B. Branches (not leaves). C. Rhizome (underground stem). D. Stem. E. Spore-producing cone. F. Whorled leaves. G. Branches and whorled leaves. This plant has silicon within its stems. The power of magnification is denoted by X.

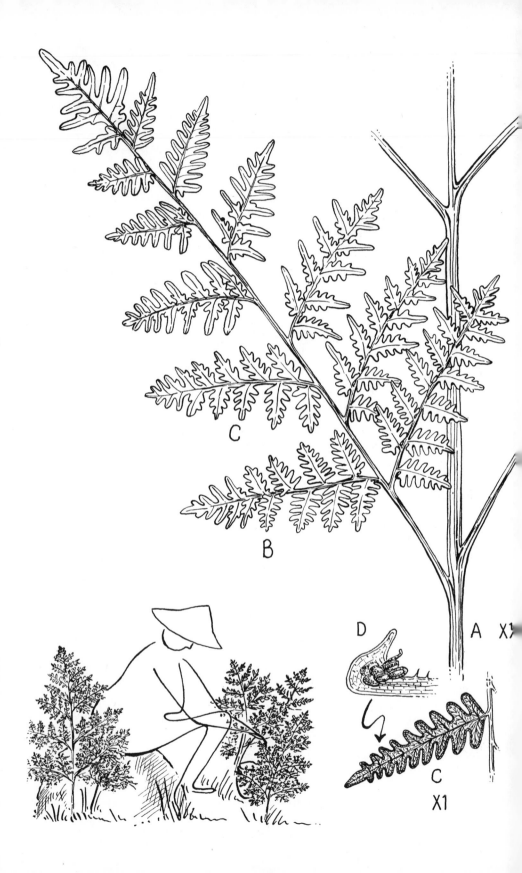

C

B

D

A X1

C

X1

Pteridium aquilinum (L.) Kuhn.

COMMON NAMES Eagle fern, bracken, brake fern, upland fern, hog brake, turkey-foot brake.

SOME FACTS Native to the United States; perennial; reproduces mainly by rhizomes, also by spores.

RANGE There are two varieties, one common to the northeastern regions of the United States and one common to the northwest; the latter is most troublesome west of the Cascades.

HABITAT Chiefly on sandy or gravelly soils in upland pastures, abandoned fields.

Few ferns fall into the category of weedy plants, but *Pteridium aquilinum*, the bracken fern, is the weediest of the ferns and is as weedy as some of our flowering plants. Few ferns can tolerate bright sunlight. They are usually plants of shady nooks, but the eagle fern is quite tolerant of bright sunlight.

The colonies of many individuals, or what appear to be individuals, are not usually the products of a sexual reproductive process but have resulted from the continued growth of the rhizome, which may be some twenty feet long. Spores are produced along the margins of the frond, and before the sporangia are mature, the margin is bent under to cover and protect them during their development. When ready to shed their spore cargoes, the margins roll forward, giving the sporangia access to the open air.

Each frond arising from the rhizome is composed of three segments, each segment being twice pinnate. The frond may be one and one-half to four feet high and Ada Georgia reports that "on the moors and mountains of Scotland the horns of the 'stately stag' are barely to be seen above it."

Boiled in salted water, the unwound croziers, which are grayish and woolly, are edible. The rootstocks have been used as a food, too, and even to make a kind of rootbeer. Ingestion by stock of the uncooked, mature fronds is a serious matter, as they contain an enzyme that destroys thiamine (vitamin B), leading to serious illnesses in cattle, sheep, and other animals. The British government aids landowners to destroy brake fern by grants of 50 percent of the cost of cutting it.

Brake fern is resistant to most chemicals, though dicamba seems to offer some hope. If you pull out the fern, take care to get the rhizome.

Eagle fern [*Pteridium aquilinum*]. A. Main rachis of fern leaf (frond). B. Pinna (leaflet), itself divided into (c) pinnules. D. The edge of the pinnule, bending to protect the sporangia (containing spores to be shed later). The power of magnification is denoted by X.

APPENDICES

A

achene A dry indehiscent (non-opening) one-seeded fruit common among the composites, though also found in other families; the sunflower "seed" is an achene, as is the true fruit of the strawberry.

alkaloid Metabolic products of plants that are nitrogen-containing bases. Some are poisonous; nicotine is an alkaloid.

alternate Any arrangement of leaves or buds on a stem in which each is placed singly and at different heights.

angiosperm The flowering plants; their seeds are "housed" in an ovary.

annual Of one season's duration from seed to maturity and death.

appressed Closely and flatly pressed against.

ascending Stems curving upward from the base.

awns A bristlelike part of appendage; found especially in the grass family.

axil Upper angle where the leaf joins the stem.

B

berry A fleshy fruit containing two or more seeds (usually more); e.g., tomato, grape, orange.

biennial Of two season's duration from seed to maturity and death; flowering usually occurs during second year.

bilaterally symmetrical Capable of division into two similiar sections that are mirror images of each other; said of a flower; e.g., snapdragon, orchid.

bipinnate Twice pinnately compound.

blade The flat, expanded portion of a leaf (as contrasted with the petiole); the photosynthetic portion of the leaf.

bract A much-reduced leaf, particularly the small scalelike leaves in a flower cluster; a small rudimentary or imperfectly developed leaf, usually green, though sometimes it is expanded and brightly colored.

bulb A bud with fleshy bracts or scales, usually subterranean; common to many members of the lily family.

C

cane The narrow flexible stem of small fruit-bearing plants; e.g., blackberry; the hollow jointed stem of taller grasses; e.g., giant reed grass.

calyx All the sepals of a flower collectively; the outer set of sterile floral leaves.

capsule A dry fruit of two or more carpels, usually dehiscent by valves splitting open along two or more divisions (sutures); e.g., the fruit of the poppy and evening primrose.

caryopsis The fruit of a member of the grass family, a grain; containing one seed fused entirely to the ovary wall. The fusion is so perfect that this fruit (ripened ovary) is popularly called a seed. A seed is a ripened ovule. Every major civilization is based on the grains or caryopses of some member of the grass family.

chlorophyll The green photosynthetic pigment found in the chloroplasts of plant cells. For photosynthesis to occur, chlorophyll must be present; thus this chemical is the source of all animal food and life.

clasping Partly or wholly surrounding the stem, as among the monocots, where the leaf lacks a petiole, though many dicots may have leaves that clasp.

composite Members of the Compositae whose inflorescence is a head or capitulum; e.g., dandelion, chicory, sunflower, or the garden aster, dahlia, daisy.

compound leaf Leaves on which the blade consists of two or more separate parts called leaflets (*see also* Pinnately compound and Palmately compound).

corm An underground stem, bulblike in outward appearance, but with a solid interior; e.g., gladiolus.

corn Once the general name for various grains (or grain plants), now, in the United States, usually limited to the corn plant (really maize plant), *Zea mays,* or to its ear of fruits; civilization in the western hemisphere has been based on this grain for thousands of years.

corolla The inner set of sterile, usually colored floral leaves; the petals taken collectively.

corymb A raceme with the lower flower stalks longer than those above so that all the flowers are seen on one level.

cotyledon Seed leaf; the primary leaf (or leaves) of a plant embryo. In the monocotyledonae there is one seed leaf; in the dicotyledonae there are two.

cruciform Having the shape of a cross, usually with four equal prongs.

culm The stem of grasses, usually hollow except at the nodes, which are swollen.

cyanogenic Giving rise to hydrogen cyanide (HCN), a deadly poisonous gas.

cyme A broad, more or less flat-topped determinate flower cluster with the central flowers opening first; most often found among the members of the pink and borage families.

D

deciduous Dying back; seasonal shedding of leaves or other structures; falling early.

decumbent Lying flat or being prostrate, with the growing tip upward.

dehiscence The splitting open of a fruit or anther to release seeds or pollen.

deltoid Triangular or delta-shaped.

dentate Toothed.

dichotomous Forking regularly in (usually) equal pairs.

digitate Diverging like the fingers of the hand.

dioecious Having male and female flowers on totally different plants.

disc (disk) flower The tubular flowers in the center of the head of many members of the composite family in contrast to the ray flowers around them; e.g., the yellow center of the daisy is composed of these flowers.

dissected Divided into many segments.

dorsal Back; back or outer surface of a part or organ.

drupe A single-seeded fleshy fruit; e.g., cherry or peach.

E

entire Used of a leaf that is neither divided nor toothed.

epidermis The outermost tissue of leaves, young roots and stems, and other plant organs.

evergreens Remaining green in all seasons. This term may be applied to flowering plants as well as to the conifers. Many members of the heath family, such as rhododendron and mountain laurel, are evergreen.

F

filament The stalk supporting the anther.

floret A single small flower of a flower cluster such as is seen in the head of the composite, or a single flower of a grass.

follicle A many-seeded dry fruit derived from a single carpel that splits longitudinally on but one side; e.g., milkweed.

frond The leaf of a fern.

G

glabrous Smooth or hairless.

glaucus Covered with bluish or white bloom (wax).

globose Spherical or nearly spherical.

glume One of the scaly bracts found at the base of and enclosing the spikelets of members of the grass family.

grain The small, hard, one-seeded fruit of a grass; e.g., wheat, rice, corn, millet, barley grains.

H

head A dense inflorescence of sessile or nearly sessile flowers on a very short axis, as seen among the composites; a capitulum. However, the flower of the clover, a legume, is also a head, though clover is not related to the composites.

herb Lacking a woody nature; a plant naturally dying to the ground in winter.

herbaceous Not woody; dying down each year.

hirsute Having rather stiff, coarse hairs.

I

imbricated Overlapping like the shingles of a roof.

imperfect (flower) Either the stamens or the carpels are lacking.

indehiscent Not regularly opening (as of seed pod or anther).

internode The part of the stem between two nodes (leaves).

inflorescence The arrangement of several flowers on a flowering shoot (as in raceme, spike, head, etc.).

L

leaflet One part of a compound leaf.

legume Dehiscent dry fruit of a simple pistil normally splitting along two sides.

lemma The outer bract surrounding the grass flower.

lanceolate Several times longer than wide; broadest near the base, narrow at the apex.

M

mechanical Said of plant tissue when it is supportive and strong.

micron One one-thousandth (1/1000) of a millimeter.

midrib The main vein of a leaf and most apparent (when apparent) in the leaf of a dicot.

monocotyledonous A plant bearing one seed leaf (or monocotyledon). Botanists shorten this to monocot. Monocotyledonous plants have parallel veins within their sword-shaped, clasping leaves.

monoecious Staminate and pistillate flowers on the same plant.

N

net-veined Said of a leaf in which the principal veins form a network, as in the dicot leaf. Such venation is also called reticulate.

node A joint or place where a leaf or leaves are attached to a stem.

O

opposite Two leaves or buds at a node; e.g., the maple or the ash.

orbicular Circular.

ovate With an outline like that of a hen's egg (broad end toward base).

ovule Undeveloped or immature unripened seed, found within ovaries of the angiosperms.

P

palea The upper of two bracts enclosing the grass flower.

palmate Radiating like a fan from approximately one point.

palmately compound Said of a leaf when the leaflets radiate from one point (usually the distal end of the petiole); e.g., the leaf of Virginia creeper.

papilionaceous (corolla) Butterfly-like in shape, as in the flower of the pea and bean; e.g., *Lotus*.

panicle An inflorescence, a branched raceme with each branching bearing a raceme of flowers, all, usually, of a pyramidal shape.

pappus A ring of fine hairs developed from the sepals covering the fruits, especially among the composites. It is the pappus that acts as the "parachute" on the small achene of the dandelion.

pedicel Stem of an individual flower in a cluster.

peduncle Stem of a solitary flower or of a flower cluster.

peltate Shield-shaped with its stalk attached within the margin.

perennial Growing many years or seasons.

perfect A flower with both stamens and pistils.

perfoliate The stem appearing to pass through the leaf.

perianth The calyx and corolla together.

petal One of the modified leaves of the corolla (in animal-pollinated flowers usually colorful).

petiole The stalk of a leaf; in celery you eat the petiole.

pinnately compound Leaflets arranged on each side of a common axis of a compound leaf; like a feather.

pistillate Having pistils and no stamens; female.

placenta The part or place in the ovary where the ovules find their attachment.

plumule The bud or growing point of an embryo plant.

pod A general term often used to designate a dry, opening fruit; e.g., a pea pod.

pollen The microspores or grains (not in the sense of fruit—*see* Caryopsis) borne by the anthers. Each grain contains the male elements or nuclei. Pollen may be animal transported or carried by the water or wind.

pome The fleshy fruit of apple, hawthorn, quince, etc.; a fruit with a bony or leathery several-celled core and a soft outer part.

prickle A small sharp outgrowth of the epidermis: e.g., the rose "thorn" is really a prickle.

privet Shrubby plants used as borders or as fencing; often members of the olive family, such as *Ligustrum*, though species of barberry (*Berberis*) may also be used. It means private and is meant to keep you that way.

precumbent Trailing or lying flat on the ground.

prostrate Lying flat on the ground.

pteridophyte A fern or related plant.

pubescent Covered with soft, fine hairs.

R

raceme A simple flower cluster of pedicled units on a common elongated axis. The flowers open from the base upward.

radial symmetry On looking down from above, the flower is circular and if any plane is passed through the center, mirror images (semicircles) are produced.

ray flower A marginal flower with a strap-shaped corolla, as seen in the composites.

receptacle The portion of the stem to which the floral organs are directly attached.

reniform Kidney-shaped.

reticulate The veins forming a network.

rhizome An underground elongated stem.

rosette A very short stem or axis bearing a dense cluster of leaves; e.g., the first year's growth of mullein.

runner A slender trailing stem taking root at the nodes when it touches the ground.

S

sagittate Arrowhead-shaped.

samara An indehiscent winged fruit as seen on the tree-of-heaven (or tree-that-grows-in-Brooklyn) in the fall.

sepal One member of the calyx.

serrate Having sharp teeth pointing forward.

sessile Lacking a petiole or stalk.

spatulate Spoon-shaped; gradually narrowing downward from a rounded summit.

spermatophyte Any seed-bearing plant.

spike A flower cluster similar to a raceme with sessile or nearly sessile flowers.

spikelet A small spike; the ultimate flower cluster in the inflorescence of a member of the grass family.

spine A sharp, more or less woody outgrowth from a leaf or a completely modified leaf; e.g., the spines seen on a barberry. (*See also* Thorn and Prickle.)

stamen The pollen-bearing or male organ of a flower.

staminate Having stamens and no pistils; male.

stellate Star-shaped.

stigma The part of the pistil that receives the pollen.

stipule A basal appendage of a petiole; usually green.

stolon A shoot that bends to the ground and takes root at the tip, giving rise to an entirely new plant.

style The part of the pistil connecting the ovary and the stigma; usually, this part is more or less elongated.

succulent Juicy; fleshy; soft and containing much water-storage tissue.

T

taproot A root system with a main root bearing smaller lateral roots; e.g., the carrot.

tendril A slender modified leaf or stem part by which a plant clings to a support.

thorn A modified, degenerated sharp-pointed branch. True thorns are found on the hawthorn, but not on the rose. (*See also* Prickle and Spine.)

tomentose Covered with dense, woollike hair.

trifoliate A compound leaf with three leaflets; e.g. clover.

tuber Usually the enlarged end of a subterranean stem. The potato is a tuber.

U

umbel An umbrellalike flower cluster found on members of the carrot family; e.g., Queen Anne's lace.

V

viscid Sticky.

ventral Front; relating to the inner face of an organ; opposite of dorsal.

W

whorls Three or more leaves or buds at a node.

wing A thin, dry membranaceous extension or flat extension of an organ; the later petals of a papilionaceous flower.

Bibliography

AGRICULTURAL RESEARCH SERVICE. 1970. The United States Department of Agriculture. *Common Weeds of the United States.* Published also as *Selected Weeds of the United States.* Washington, D.C.: United States Government Printing Office.

AMES, OAKES. 1939. *Economic Annuals and Human Cultures.* Cambridge, Mass.: Botanical Museum of Harvard University.

BAILEY, L. H. 1963. *How Plants Get Their Names.* New York: Dover Publications.

COCANNOUR, JOSEPH A. 1950. *Weeds, Guardians of the Soil.* New York: Devin Adair Company.

CHANCELLOR, R. J. 1966. *The Identification of Weed Seedlings of Farm and Garden.* Oxford: Blackwell Scientific Publications.

CRAFTS, ALDEN S. 1975. *Modern Weed Control.* Berkeley: University of California Press.

CRONQUIST, ARTHUR. 1968. *The Evolution and Classification of the Flowering Plants.* New York: Houghton Mifflin.

EMERSON, BARBARA H., ED. 1965. *Handbook on Weed Control.* New York: Brooklyn Botanical Garden.

ESAU, KATHERINE. 1960. *Anatomy of Seed Plants.* New York: John Wiley & Sons.

FERNALD, MERRITT LYNDON. 1950. *Gray's Manual of Botany.* 8th ed. New York: American Book Co.

FERNALD, MERRITT; KINSEY, LYNDON; AND KINSEY, ALFRED CHARLES. 1958. Revised by Reed C. Rollins. *Edible Wild Plants of Eastern North America.* New York: Harper & Row.

FOGG, JOHN M., JR. 1945. *Weeds of Lawn and Garden.* Philadelphia: University of Pennsylvania Press.

GEORGIA, ADA E. 1942. *A Manual of Weeds.* New York: The Macmillan Co.

GIBBONS, EUELL. 1966. *Stalking the Healthful Herbs.* New York: David Mackay.

―――. 1965. *Stalking the Wild Asparagus.* New York: David Mackay.

GLEASON, HENRY A. 1952. *The New Britton and Brown Illustrated Flora of the Northeastern United States and Adjacent Canada.* Lancaster, Pa.: New York Botanical Garden.

―――. 1947. *Plants of the Vicinity of New York.* New York: New York Botanical Garden.

HEDRICK, U. P., ED. 1972. *Sturtevant's Edible Plants of the World.* New York: Dover Publications.

HITCHCOCK, A. S. 1971. Edited by Agnes Chase. 2 vols. *Manual of the Grasses of the United States.* New York: Dover Publications.

HOLM, LEROY. 1971. *The Role of Weeds in Human Affairs. Weed Science* 19: 485–490.

JAQUES, H. E. 1959. *How to Know the Weeds.* Dubuque, Iowa: William C. Brown.

KALM, PETER. 1966. *Travels in North America.* 2 vols. New York: Dover.

KING, LAWRENCE J. 1966. *Weeds of the World.* New York: Interscience.

KINGSBURY, JOHN M. 1965. *Deadly Harvest.* New York: Holt, Rinehart and Winston.

————. 1964. *Poisonous Plants of the United States and Canada*. Englewood Cliffs, N.J.: Prentice Hall.

KIRK, DONALD. 1970. *Wild Edible Plants of the Western United States*. Healdsburg, Cal.: Naturegraph.

LAWRENCE, GEORGE H. M. 1958. *Taxonomy of the Vascular Plants*. New York: The Macmillan Co.

MILLER, JAMES. 1969. *Weed Identification*. Bulletin 632. Athens, Ga.: University of Georgia College of Agriculture.

MONTGOMERY, F. H. 1964. *Weeds of the Northern United States and Canada*. New York: Frederick Warner & Co.

MEUNSCHER, WALTER CONRAD. 1960. *Weeds*. 2nd ed. New York: The Macmillan Co.

SPENDER, E. R. 1954. *Just Weeds*. New York: Charles Scribner's Sons.

SMITH, A. W. 1963. *A Gardner's Book of Plant Names*. New York. Harper & Row.

SALISBURY, SIR EDWARD. 1964. *Weeds and Aliens*. London: Collins.

UNITED STATES DEPARTMENT OF AGRICULTURE. 1948. *Grass, the Yearbook of Agriculture*. Washington, D.C.: United States Government Printing Office.

————. 1951. *Seeds, the Yearbook of Agriculture*. Washington, D.C.: United States Government Printing Office.

WHEELWRIGHT, EDITH GREY. 1974. *Medical Plants and Their History*. New York: Dover.

INDEX

wild cotton, 233
wild cucumber, 9, 20, 59
Wilde, Oscar, 121
wild endive, 223
wild garlic, 8, 73
wild grape, 55–56
wild hemp, 117
wild lady's slipper, 171
wild millet, 79
wild morning glory, 19, 51
wild onion, 20, 73
wild parsnip, 25, 163
wild phlox, 213
wild portulaca, 129
wild snapdragon, 137
wild strawberry, 26, 181
wild sunflower, 153
wild sweet william, 213
wild tansy, 119
wild tomato, 189
wild vetch, 63
willow herb, 239
wineberry, 47
wine's grass, 175
winter annuals, 5
winter cress, 155
winterweed, 179

wire grass, 75, 87
witloof, 223
wolf grass, 83
wonderberry, 189
woodbine, 65
wood phlox, 213
wormwood, 23, 121

Xanthium pennsylvanicum L., 22, 115

yard grass, 87
yard rush, 75
yarrow, 28, 209
yellow daisy, 169
yellow devil, 133
yellow dock, 123
yellow-flowered species, 18, 21, 23–25,
    129–67
yellow foxtail grass, 79
yellow-green–flowered species, 18, 21,
    23, 123–27
yellow myrtle, 131
yellow paintbrush, 133
yellow rocket, 24, 155
yellow sweet clover, 211
yellow toadflax, 139